T0146179

Geographies of Knowledge

Medicine, Science, and Religion
in Historical Context
Ronald L. Numbers, *Consulting Editor*

GEOGRAPHIES OF KNOWLEDGE

SCIENCE, SCALE, AND SPATIALITY

in the

NINETEENTH CENTURY

EDITED BY

Robert J. Mayhew

&

Charles W. J. Withers

Johns Hopkins University Press

Baltimore

Printed in the United States of America on acid-free paper
2 4 6 8 9 7 5 3 1

Johns Hopkins University Press
2715 North Charles Street
Baltimore, Maryland 21218-4363
www.press.jhu.edu

Library of Congress Cataloging-in-Publication Data

Names: Mayhew, Robert J. (Robert John), 1971– editor. |
Withers, Charles W. J., editor.
Title: Geographies of knowledge : science, scale, and spatiality in the nineteenth century / edited by Robert J. Mayhew and Charles W. J. Withers.
Description: Baltimore : Johns Hopkins University Press, [2020] | Series: Medicine, science, and religion in historical context | Includes bibliographical references and index.
Identifiers: LCCN 2019047623 | ISBN 9781421438542 (hardcover) |
ISBN 9781421438559 (ebook)
Subjects: LCSH: Science—History—19th century. | Scientific expeditions—History—19th century. | Intellectual life—History.
Classification: LCC Q125 .G454 2020 | DDC 509/.034—dc23
LC record available at https://lccn.loc.gov/2019047623

A catalog record for this book is available from the British Library.

Special discounts are available for bulk purchases of this book. For more information, please contact Special Sales at specialsales@press.jhu.edu.

Johns Hopkins University Press uses environmentally friendly book materials, including recycled text paper that is composed of at least 30 percent post-consumer waste, whenever possible.

CONTENTS

JOHN A. AGNEW is Distinguished Professor of Geography at the University of California–Los Angeles. The author of numerous books, his research interests include political geography, international political economy, and questions about the ways in which geography can figure in the origins and circulation of ideas and theories. Among his most notable books are *Place and Politics: The Geographical Mediation of State and Society* (1987), *Geopolitics: Re-visioning World Politics* (2nd ed., 2003), and (as coeditor) *The SAGE Handbook of Geographical Knowledge* (2011) and the *Wiley-Blackwell Companion to Human Geography* (2011).

VINITA DAMODARAN is professor of South Asian history and director of the Centre for World Environmental History at the University of Sussex. Her work on the environmental history of South Asia includes the use of historical records to understand climate change in the Indian Ocean world. Her publications include *Nature and the Orient* (2000; coedited with Richard Grove and Satpal Sangwan), *British Empire and the Natural World: Environmental Encounters in South Asia* (2010; coedited with Deepak Kumar and Rohan d'Souza), *East India Company and the Natural World* (2014; coedited with Anna Winterbottom and Alan Lester), and *Climate Change and the Humanities* (2017; coedited with Alexander Elliott and James Cullis).

DIARMID A. FINNEGAN is senior lecturer in human geography at the Queen's University, Belfast. He is the author of *Natural History Societies and Civic Culture in Victorian Scotland* (2009), which won the Frank Watson Book Prize for Scottish History. His research is concerned with science, space, and culture in historical perspective, including a project on science in nineteenth-century Belfast, funded by the Arts and Humanities Research Council, as well as a work in progress on the reception of ideas about human evolution in the context of religious debates about the creation of Eve and on geography and Christian missions in late Victorian Britain.

NUALA C. JOHNSON is professor of geography at Queen's University, Belfast. She has written extensively on nationalism and the politics of identity, public monuments and collective memory, literary spaces, and the historical geographies of science. She is author of *Ireland, the Great War and the Geography of Remembrance* (2003) and *Nature Displaced, Nature Displayed: Order and Beauty in Botanical Gardens* (2011), and has edited three further books. She is currently working on a project on gender, empire, and natural history in colonial Burma.

DANE KENNEDY is Elmer Louis Kayser Professor of History and International Affairs at George Washington University. His books include *The Imperial History Wars: Debating the British Empire* (2018), *The Last Blank Spaces: Exploring Africa and Australia* (2013), *The Highly Civilized Man: Richard Burton and the Victorian World* (2005), and, as editor, *Reinterpreting Exploration: The West in the World* (2014). For the past five years, he has also directed the National History Center in Washington, DC.

ROBERT J. MAYHEW is professor of historical geography and intellectual history at the University of Bristol. He is the author of *Enlightenment Geography* (2000), *Landscape, Literature and English Religious Culture* (2004), and *Malthus: The Life and Legacies of an Untimely Prophet* (2014). He was a Philip Leverhulme Prize winner for his work in historical geography and is a fellow of the Society of Antiquaries of London. He has edited Malthus's works for Penguin Classics (2015) and *New Perspectives on Malthus* for Cambridge University Press to commemorate the 250th anniversary of Malthus's birth in 2016.

MARK NOLL is Francis A. McAnaney Professor of History Emeritus at the University of Notre Dame. His books include *America's God, from Jonathan Edwards to Abraham Lincoln* (2002), *The Civil War as a Theological Crisis* (2006), *God and Race in American Politics: A Short History* (2008), and *In the Beginning Was the Word: The Bible in American Public Life, 1492–1783* (2016). With David Livingstone he edited *Charles Hodge's What Is Darwinism? And Other Writings on Religion and Science* (1994), *Evangelicals and Science in Historical Perspective* (1999), and *B. B. Warfield: Evolution, Science, and Scripture—Selected Writings* (2000).

RONALD L. NUMBERS is Hilldale Professor of the History of Science and Medicine Emeritus at the University of Wisconsin–Madison. A fellow of the American Academy of Arts and Sciences and a recipient of the Sarton

Medal for lifetime achievement in the history of science, he is the author or editor of thirty books, including *Darwinism Comes to America* (1998), *Disseminating Darwinism* (coedited, 1999), and *The Creationists* (expanded edition, 2006). In retirement, he is writing a biography of John Harvey Kellogg (for Harvard University Press).

NICOLAAS RUPKE is Johnson Professor of History at Washington and Lee University. He is the author of *The Great Chain of History: William Buckland and the English School of Geology* (1983), *Richard Owen: Biology without Darwin* (1994), *Alexander von Humboldt: A Metabiography* (2005), and numerous other books. His research interests span the history of science and medicine, national science of Great Britain and Germany, science and society, science and religion, history of evolutionary biology, and scientific biography. Current research projects address the non-Darwinian tradition in evolutionary biology and a scientific biography of Johann Friedrich Blumenbach.

YVONNE SHERRATT is senior lecturer in the School of Geographical Sciences at the University of Bristol. Educated at the University of Cambridge, she has taught at Cambridge, Oxford, Edinburgh, and Wales. She is the author of *Adorno's Positive Dialectic* (2002), *Continental Philosophy of the Social Sciences* (2005), and, most recently, *Hitler's Philosophers* (2013), the last of which has been translated into nine languages. Her current work focuses on philosophy, music, and exile in the mid-twentieth century.

CHARLES W. J. WITHERS is professor emeritus in geography at the University of Edinburgh. His books include *Geography, Science, and National Identity: Scotland since 1520* (2001), *Placing the Enlightenment: Thinking Geographically about the Age of Reason* (2007), *Geography and Science in Britain, 1831–1939* (2010), *Zero Degrees: Geographies of the Prime Meridian* (2017), and, with Innes Keighren and Bill Bell, *Travels into Print: Exploration, Writing, and Publishing with the House of Murray, 1773–1859* (2015). His edited books include *Geography and Enlightenment* (1999), *Geography and Revolution* (2005), and *Geographies of Nineteenth-Century Science* (2011), all with David Livingstone, and, with Fraser MacDonald, *Geography, Technology and Instruments of Exploration* (2015).

PREFACE

Over the past thirty years or so, modern scholars have not only questioned historical presumptions about the rise of modern science in the nineteenth century but also turned strongly to geographical thinking to explain the making, reception, and mobility of science and scientific cultures in that period. This volume is a contribution to continuing debate in this context and, we hope, a stimulus to further work. The chapters examine science's "spatialization" by exploring the ways in which geography, place, and, importantly, scale inflected the production, circulation, and reception of ideas in the long nineteenth century.

We are grateful to our authors for providing us with such engaging studies and for their forbearance of us as editors in bringing the material together. It has been a pleasure to work with such a collection of authors. As will be clear from the individual chapters and, we hope, from our introduction, this book is an acknowledgment of the intellectual debt and stimulus we and others have all been afforded by the scholarship of David Livingstone on the geographical examination of science, religion, and aspects of intellectual history, chiefly but not alone in the nineteenth century. Each of us has had the good fortune at one time or another and sometimes more than once to write with him, or to be a colleague, or to have benefited from his intellectual largesse and scholarly support warmly given.

If the book, then, is recognition of the man and of his work, it is written in the hope that these chapters will advance still further what now can be thought of as the geographies of science—and for periods other than the nineteenth century. There is every reason to claim that what has become known as late twentieth-century "technoscience," or the concerns of seventeenth- and eighteenth-century natural philosophers, can be similarly illuminated by reference to the places in which these topics were made, the processes by which the findings moved, and the practices of representation through which they were codified. For the twentieth century especially perhaps, the connections among science, politics, and civil society lend themselves to geographical interpretation: made in specific venues, often disputed nationally yet shared in international context even as publics everywhere have lost sight of increasingly specialist discourses, if not also the implications of science itself.

We acknowledge with grateful thanks the afterword by John A. Agnew as well as his kindness in reading through all the contributions here. It has been a pleasure to work with the Johns Hopkins University Press editorial and production team, and we thank them for their courteous and timely support. The two readers of the proposal on which this collection is based are owed thanks for their insightful comments and their suggestions for improvement, which we have been pleased to accept. For the illustrations that form part of Nuala C. Johnson's chapter, we are grateful to the National Botanic Gardens, Glasnevin. If our thanks are due, as they are, to our authors here, we cannot let pass this opportunity to acknowledge our respective colleagues in Edinburgh and in Bristol for their support. At times of late, working in the academy has been fraught and dispiriting. Having the collective and collegial support of colleagues and authors and a university press who value thinking and writing in the ways they do has been, and continues to be, a great fillip.

Geographies of Knowledge

Thinking Geographically about Science in the Nineteenth Century

ROBERT J. MAYHEW AND CHARLES W. J. WITHERS

In a review in 2008 of "ten problems in history and philosophy of science," historian of science Peter Galison addressed (as "problems 7 and 8") the topics of "locality and globality." Of the first, he noted that "the turn toward local explanation in the historical, sociological, and philosophical understanding of science may well be the single most important change in the last thirty years."[1] For Galison, notions of locality entail understanding "the microhistorical enterprise," by which he meant those "subtle interconnections of procedures, values, and symbols that mark science in a place and time, not as a method but as a kind of scientific culture." What kind of explanation, Galison asked, "is involved in the microhistorical enterprise?" Microhistory—and, by implication, we might add "microgeography"—"is supposed to be exemplification, a display through *particular* detail of something *general*, something more than itself." This led to a hard question: "What does it mean to aim for exemplification without typicality?" In contrast, globality embraced those aspects of scientific practice that, in Galison's view, "simply do not reduce to the local." Whatever the "larger, normative roles, techniques, and methods" under consideration, he cautioned, there is a danger in looking at such matters of science's making in too much detail: "They are invisible when the view is too close." So, as Galison put it: "To the problem, then: What aspects of scientific practice, scientific argumentation, and scientific self are not visible when looked at by a microstudy? Why?"[2]

A considerable body of work in the history and philosophy of science exists to support Galison's declaration and to illustrate and elaborate upon these twin specific concerns. As that work makes clear, this "single most important change" is neither simply nor only a "turn toward local explanation." It can be characterized more broadly as a turn toward the importance of place and space as explanatory categories in the study of science in historical and contemporary context. This is itself part of a wider "spatial turn" in the humanities and social sciences that recognizes that "*where* things happen is crucial to knowing *how* and *why* they happen."[3] Thirty years ago, to speak at all of the spatial nature of science, never mind its local or global dimensions, would have been to risk imprecision in the utterance and bafflement in one's audience. Today, it is possible to speak of the geography of science and for that term to convey the importance of studying science's spatial expression; its situated contexts of discovery and justification; its local, regional, national, and global dimensions; and science's mobility as a way of knowing and a form of communication between and among practitioners and their audiences.

Much of this work has been the subject of review elsewhere. It is not our purpose here to address it in any detail, but a few remarks upon its main features are pertinent. Some scholars have turned to address in general terms the significance of space and place in the making and reception of science, while others have proposed particular regional or thematic typologies in thinking about science's geographies. The article by Adir Ophir and Steven Shapin in 1991 and, in separate publications, the work in 1998 of Golinski, Shapin, Harris, and Smith and Agar were each formative influences in this respect.[4] More recently, and as studies of particular sites in science continued, others scrutinized the situated nature of knowledge more generally: in both senses, geography's lexicon of place, space, scale, site, and so on affords powerful ways to explain, to again cite Galison, the "subtle interconnections of procedures, values, and symbols that mark science in a place and time."[5]

This is, of necessity, a greatly truncated intellectual genealogy of the past emergence and present strength of the geographical study of science in historical context. It cannot be allowed to pass without noting, however, the central place occupied by geographer of science and of religion David Livingstone, who, over the course of the past thirty years or so, has imbued significantly those *what*, *how*, *who*, and *why* questions intrinsic to the historical study of science with the importance of *where*. The present volume emerged as a project from the shared desire of a group of scholars to engage with and reflect upon the totality of Livingstone's contribution to our understanding of the intellectual conjunctions of science and spatiality as a starting point for new insights into the histories of scientific endeavors. Where Livingstone's work has obviously stretched across the broad chronological swath of postmedieval science, it was decided that the contributors would cluster their engagements with science and spatiality around the "long" nineteenth century, which has been at the heart of his work on religion, variants of Darwinism, and the preadamite controversy. Our ambition, then, is not to create a celebratory Festschrift but to use Livingstone's work as the originating focus from which new engagements and understandings of the spatiality of science will emerge—engagements and understandings that further and challenge his work as well as simply respond to it.

The influence of Livingstone's work in shaping the geographical study of science and in advancing spatialized approaches to account for the performance and practice of scientific culture is clear in the chapters that follow. Here, too, and for reasons of space, our purpose is not to dwell upon either the detail or the influence of specific publications: for the most part, our authors do that. Rather, it is to suggest that we can discern in Livingstone's work generally a shift from those early and perhaps hesitant declarations concerning the importance of space, place, and geographical mobility in explanation of science's history to, later, greater confidence in, and refinement of, those self-same geographical approaches.[6] This confidence is evident in statements about the scales at which such approaches might proceed and in substantive works of review and scholarship. Such an interpretation would be justified, we suggest, of Livingstone's 1994 paper, wherein he spoke of "the value of inserting spatiality into the enterprise traditionally known as 'the history of ideas' and thus reconceptualising it as 'the historical *geography* of ideas.'"[7] So, too, his 1995 paper, in which he proposed a threefold schema for the historical geography of science (regionalization of scientific style, political topography of scientific commitment, social space of scientific sites), put forward a "rudimentary agenda" for the historical geography of geography and emphasized the importance of taking seriously "the situatedness of knowing" in geography's history.[8] This is also the case for his 2003 book *Putting Science in Its Place*, important papers about reading and speech in the shaping and dissemination of science, and later monographs on race and religion as well as the geography of Darwinism.[9]

Rather than being just part of the spatial turn, Livingstone's has been a leading voice in elucidating what the geography of science encompasses and why it matters. In particular, his work has chronologically centered its attention on interpreting the spatial dimensions necessary to understand those complex relationships in the nineteenth century between proponents of science and religion, as well as their often conflicting ideological, political, and personal motivations. In what follows, his work is drawn upon in different ways and with respect to different places and scales.

Obviously, David Livingstone has no monopoly on insights into the conjunctions of science and spatiality in the nineteenth century. As will become clear both in this introduction and in the chapters that follow, our intention in using Livingstone's work as a touchstone is not to deny the seminal contributions of others but merely to provide a coherent tie for the thematic and conceptual concerns of our contributors. In seeking to take the dialogue beyond the point to which Livingstone and others have taken it, a key conceptual feature of the present collection of chapters is its attention to scale. We will say more about the debates about scale that have galvanized contemporary human geography later in this chapter. For all the conceptual sophistication and empirical richness of the conjunctions of science and spatiality that students of the history of nineteenth-century science have exhibited, they have worked through the rubrics of "space" and "place," the former as a more abstract term for a territorial extent and the latter as the lived reality of that territory as experienced by scientific communities. While other spatialized terms have been deployed by historians of science—notably "networks" and "migration/diffusion"—scale has been notable for its neglect as a spatial category of analysis. Bernard Lightman's seminal collection *Victorian Science in Context* (1997), for example, is more concerned with the emergence of modern demarcations of sciences and their boundary disputes. While the final section of that book, titled "Practising Science," clearly attends to various sites and to different scales from the local to the global, the scalar itself is not a category that the essays seek to problematize. The attention to epistemological rather than spatial territories and their borders is even more pronounced in the essays collected by David Cahan in *From Natural Philosophy to the Sciences* (2003). Similar comments hold true for James A. Secord's *Victorian Sensation* (2000), whose section titled "Geographies of Reading" attends in detail to the local scale at which readings of Robert Chambers's *Vestiges of the Natural History of Creation* were built and to the differences that different spaces made to those readings, but it does not seek to move across scales to determine how that affected interpretations of the text. The most directly comparable venture to the present one is Livingstone and Withers's *Geographies of Nineteenth-Century Science* (2011). The first section of the Livingstone and Withers volume attends to "Sites and Scales," but in the course of five chapters the ability to cover the range of scales from local to global is inevitably more constrained than that in the present volume. Further, it is only really Withers's chapter on the activities of the British Association for the Advancement of Science that deploys scale as a conceptual category, and that attends to the emerging culture of civic science in the towns of Victorian Britain. In short, the essays in Livingstone and Withers offer an important source of inspiration for the scalar rationale of the present collection, but in attending as they do mainly to the local and by not problematizing scale as a category they leave much work to be done in conjoining science and scale in the nineteenth century for the contributors to the present volume.[10]

We are not claiming, of course, that our book is the first attempt to consider issues of scale in the history of science, but the interest in scale as a conceptual category has only emerged during the gestation of this project. Perhaps most importantly, Deborah Coen's *Climate in Motion* (2018) interrogates the ways in which the emergent inquiry of climate science had to wrestle with the interleaving of different scales of analysis from the mid-nineteenth century, particularly attending to the ways in which that science operated in the territories of Central Europe and sought to relate local weather and climate to emerging notions of a general circulatory atmospheric system. One of the main problematics that defined climate science in its early phases (and, of course, still today), then, was its need to integrate analyses at different scales. In a similar vein, Nathan Sayre's analysis of the tradition of rangeland science in the United States, *The Politics of Scale* (2017), stretches back to the later nineteenth century and shows the issues scientists and managers of physical environments had in "upscaling" concepts from the emergent discourses of biology and ecology to the vast rangeland spaces they sought to preserve and utilize. Finally, Lydia Barnett has recently suggested that this scientific concern with the scalar has a longer trajectory than Coen's and Sayre's analyses alone can disclose. She suggests that European discourses about deleterious anthropogenic impact on the global environment stretch back to the later Renaissance and were deeply concerned with how purportedly global events such as Noah's Flood were evidenced by local scientific inquiries and vice versa.[11]

The contributors to this volume address scale, space, and science in a variety of ways. In the first of two chapters examining questions of "locale," Robert J. Mayhew and Yvonne Sherratt provide a spatially inflected textual reading of Thomas Malthus's hugely influential 1798 *Essay on the Principle of Population*, notably the expanded "Great Quarto" edition of 1803. Where others position Malthus chronologically in the canon of influential political economists, Mayhew and Sherratt employ a spatial hermeneutics to interpret his work locationally. Diarmid A. Finnegan takes us to Belfast (for long, Livingstone's own academic milieu), whose citizens on three separate occasions, in 1852, 1874, 1890, and for different religious and moral reasons, variously reviled or roared on the speeches and writings of physicist John Tyndall.

Then follow three chapters whose central focus is at the scale of the nation. Mark Noll displays a similar "sensitivity to *the spaces of a life*" as Finnegan does of Tyndall in Belfast in his account of American proslavery and racist thinker Henry Hotze.[12] Place matters in life writing: "reputational geographies" vary over space as well as over time.[13] Ronald L. Numbers's account, hardly at all biographical in focus, considers the settings and textual disputes over race and religion in the nineteenth-century United States, and, later and elsewhere, in anti-evolution and anti-Darwinian debates in America, Britain, Australia, and New Zealand. In his study of the structuralist theory of evolution, principally but not alone with reference to Germany, Nicolaas Rupke reminds us that while specific scientific beliefs in the nineteenth century could and did have determining antecedents, so could they also have a lingering "afterlife," even "afterlives," within given nations and across national boundaries.

The final four chapters turn to global matters. Charles W. J. Withers's chapter on the prime meridian in late nineteenth-century international geographical conferences exemplifies those tensions and connections between locality and globality identified by Galison: how a matter of global regulation and international disputation took shape through iterative negotiation in specific and short-lived scientific settings. Nuala C. Johnson's chapter on the botanical collecting and representational practices of Charlotte Wheeler-Cuffe in Burma highlights her "in-the-field" search for specimens "out there" in nature, and the fact that botanical science could take shape in domestic spaces and at a rhythm influenced by others. Some sciences were pursued in places and in ways that blur the distinction between "laboratory" and "field," or, even, "home" and "garden": botany and biology are illustrative in this latter respect, as the experiences and practices of Wheeler-Cuffe and other examples show.[14] So, too, to give another example, was meteorology, whose concern with measurement and the spatially distributed "laws" of nature driving variations in weather was shaped variously by work in country-house gardens, British naval vessels, and imperial observatories.[15]

Precisely because science would come to have international, continental, or global dimensions—in what it studied and in where it was studied—it behooves us to examine the performance of science in place *and* in its mobility across global space.[16] It may matter in this regard not just what that science or particular discourse focused upon but also the evident consequences of the work in question. As Vinita Damadoran shows in assessing the connections nineteenth-century contemporaries made between deforestation and desiccation, the emergent environmental sciences had peculiarly colonial origins that were "exported" to institutions at the metropole. Changes to the climate in Britain's colonies had, it was argued, lessons for the empire—and the world—as a whole. Similarly, overseas exploration and expeditionary culture could and did associate science with imperial politics. Those politics, while important, were often overlooked in the face of unknown or adversarial environments or as the result of lack of forethought by administrators and scientists alike. As Dane Kennedy shows in his account of expeditionary endeavor in Africa, neither the production in advance of textual guides nor the advice in situ of human guides necessarily prepared traveling scientists to "do science" and safely return, specimens and notebooks in hand, unsullied by the rigors of firsthand empirical encounter with the unknown.[17] Global science in this sense—exploration-cum-expeditionary culture—was something altogether different from the experiences of Charlotte Wheeler-Cuffe or the polite polyvocal debate in metropolitan meetings designed to regulate the world's metrology.[18]

We return in the following section to the concerns of our individual authors and to the connections among their essays, on both of which John A. Agnew's afterword offers an elegant and insightful commentary. It will be clear, too, that the ordering principles of "locale," "national," and "global" here are helpful, but they are not fixed categorizations. Questions of site and scale—key elements in the spatiality of science—have the utility they do not because they are fixed in their meaning, but because they are fluid. Being so, they invite us to consider the forms science takes as it is shaped in certain venues, and as it moves among them as something embodied and, at least potentially, at once local, national, and global. Before turning to these issues, we consider the wider context of science in the nineteenth century to observe that narratives about science in general or about the period in question should not obscure differences between and among the sciences or variations over time in what, where, and how they worked.

Space, Place, and Spatiality in Nineteenth-Century Science

Whatever else science was or would become in the nineteenth century—a form of cultural capital and political power, a profession, discipline, a preeminent means to explore and explain the natural world or to interpret and manage society—modern interpretations should not lose sight of the fact that, for contemporaries, much was new, in method, subject, and direction. If science was natural philosophy made particular and, at least initially, was without agreed-upon ends in view, there is no doubt that it became profoundly transformative, for participants and audiences alike. For historian of science David Cahan, "developments in the sciences in this period arguably equalled or exceeded those in natural philosophy during the Scientific Revolution of the sixteenth and seventeenth centuries, and in virtually every respect, be it intellectual range, theory formation, empirical results, or instrumentation. Moreover, the sciences underwent unprecedented institutional growth and had a large role in reshaping society—just as society helped reshape them."[19] The emergence of science in these terms and the shaping of individual disciplines during the 1800s was neither abrupt nor everywhere the same, but the facts of change and their consequences were not in doubt.

> Certainly by the final third of the nineteenth century, one could speak legitimately, that is, in a modern sense, of "science," "scientists," and the disciplines of science. These new labels and categories reflected the fact that science had both delimited itself more fully from philosophy, theology, and other types of traditional learning and culture and differentiated itself internally into increasingly specialized regions of knowledge. At the same time, new institutions, such as specialized societies and institutes, were created, and the notion of a "scientific community" appeared. Moreover, interactions between and among the sciences and other aspects of culture, the economy, the state, and society in general became more significant. In many minds "the nineteenth century" and "science" became synonymous with "progress."[20]

These observations are mirrored for social science in the same era, which crystallized its identity and subdivisions as well as clarified both its indebtedness to and distinction from the natural sciences in discernibly "modern" ways.[21]

As Cahan further notes, different explanations have been advanced to account for these developments. Several earlier modern accounts attempt a "grand narrative" approach. In his *A History of European Thought in the Nineteenth Century* (1904–12), for example, English industrial chemist John Merz rested his explanation upon what he termed the rise of "scientific spirits" and different analytic "views of nature"—astronomical, morphological, statistical, and so on. English crystallographer John Bernal offered a broadly Marxist interpretation in his *Science in History* (1954), arguing that science in the 1800s owed its emergent authority to material needs consequent upon the growth of urban-industrial capitalism. In his *The Scientist's Role in Society* (1971), Israeli-American historical sociologist Joseph Ben-David interpreted science's rise to prominence as the result of state patronage and capitalism's ethic of free enterprise and individual competition.[22]

It is the case, as Cahan has also observed, that scholarship has more recently shied away from "big picture" explanation—a move consistent, of course, with Galison's observations about the turn toward local explanation. What Cahan has called "the social reconfiguration of the sciences during the nineteenth century" is distinguished more by studies of leading men and women of science; the shaping of particular disciplines; the relations between medicine, science, and technology; and attention to institutional and political circumstances.[23] In addition to the "defining" features of subject specialization, institutional formation, and closer links to the state, science in the nineteenth century was concerned more than in earlier periods with the implications of standardization, with measurement, with discerning natural and social laws, and with explanations rooted in theory rather than in unexaminable belief and the rhetorical appeal of religious dogma. Science and those who undertook it became more professional in terms of personnel, published outputs, in their institutional settings, and so forth.[24] In short, in the nineteenth century science constantly sought new grounds on which to operate—novel matter and places of concern—even as it shaped itself. Because this is true, "the historiography of nineteenth-century science" as Cahan has further observed, "is still very much a work in progress."[25]

So, too, what we might think of as the "geo-historiography" of nineteenth-century science—the study of science's locatable dimensions, not just its datable ones—continues as a work in progress, as the evidence of the past thirty years or so testifies.[26] Because this is so, we should always be looking to further questions concerning science's specific spatialities in this period, mindful of science's general historical origination. One such question might be to disclose contemporary recognition of the importance of spatial variation within the sciences. How far did nineteenth-century practitioners recognize *where* questions as of fundamental importance to the emergence of their subject? Take the botanical sciences as an illustration. Where plants were in the world in the historical present, and the general study of biogeographical origins of which botanical classification and mapping were a part, were clues to Earth's history. What adherents to divine authority read as fixed and preordained, natural scientists increasingly interpreted as fluid and evolutionary. The botanical realm saw a recognition of the role of geography from early in the nineteenth century thanks to the impact of Alexander von Humboldt's *Essai sur la Géographie des Plantes* (1807). In nineteenth-century Germany, for example, the scientific study of plants, either as "botanical arithmetic" (in German *Tabellenstatistick*, the number of species in an area) or the analysis of species and taxa in relation to soil, altitude, and climate by floristic "province," depended upon questions of distribution to represent findings and to explain relationships between plants and other living things.[27]

In Britain, botanist Joseph Dalton Hooker made just such points in his address to the geographical section of the British Association for the Advancement of Science in 1881: "In the science of distribution, Botany took the lead." He went on to note that "under the theory of modification of species after migration and isolation," here referring to Darwin's theory of natural selection as outlined in his *On the Origin of Species* (1859), "their representation in distant localities is only a question of time and changed physical conditions." What Darwin earlier described to Hooker as "that almost keystone of the laws of Creation, Geographical Distribution," Hooker took to be historical truth, observable fact, and a template for future scientific study: "With the establishment of the doctrine of the orderly evolution of species under known laws, I close this list of those recognised principles of the science of geographical distribution, which must guide all who enter upon its pursuit. As Humboldt [polymath Alexander von Humboldt] was its founder, . . . so we must regard Darwin as its latest and greatest lawgiver."[28] As botany in the nineteenth century would become the science of phytogeography in its object of study, it was with other sciences also spatial in its practices of representation. Mapping living plants—as also birds and animals and, in ethnology, human variations—allowed scientists to think of regional patterns and distributional pathways in the natural world as dynamic rather than static and, from that, to explain distribution over time and space.[29] Whether all scientific inquiries beyond the life sciences were equally alert to geography is a moot point, but we can say with confidence that from long before publication of William Smith's pioneering geological map of Great Britain in 1815, geology and the earth sciences were thus alert.[30]

If science in the nineteenth century was "emergent," becoming disciplined and so on over time, we must similarly allow that it may have done so differently over space, and so, where we can, we must show how and why it varied in its spatial settings and in its consequences. It may be, indeed, that the nineteenth century is especially important in this respect precisely because science was "in formation" and because specific disciplinary practices emerged in this period. Whether the nascent discipline was botany; geology with its interests in space, "deep time," and inferential reasoning over what lay beneath Earth's surface; or those aspects of physics, chemistry, or biology that depended upon experimental representation of the invisible, science in its conduct and in its reception took place somewhere.

What matters is to show how and why science was thus conditioned, in its epistemic practices and social cultures, by the venues in which it was undertaken and received. Almost a decade ago, the claim was made that the various classificatory schema for thinking spatially about science were, at best, "rough approximations for thinking about two compelling sets of questions: the first having to do with the making and meaning of science *in place*, the second with science's movement *over space*."[31] Nothing over the past ten (or thirty) years or so about work in the geography of science suggests that the twin geographical concerns of place and space have lost any utility through continued critical application in different contexts. Quite the opposite: their currency, if anything, is enhanced. It is arguable, of course, that used in the abstract, in a strictly nominalist sense, the categories of "place" and "space" as key elements of "the rubric of the historical geography of science" may be not much more than powerful "spatial metaphors." It is equally the case that these and other terms have been used—and should continue to be used—in evidently realist and material senses—that is, that the spaces and places of science's production and reception are and were real entities: socially constitutive, empirically active, discursively productive.[32] It is therefore important, given the currency of science read spatially, that further questions follow "beyond" those of place and space and with more specific purchase.

A host of questions emerge. Given the general emergence of science in the nineteenth century, what do the geographical dimensions of different sciences look like? What were the specific material forms in and through which increasingly individuated sciences were put to work and made to travel: visually, in teaching texts, in experimental regimes, in speech spaces, in public review? What does it mean to consider the geography of science in general, or of particular sciences, at different scales? In a sense, this last question is perhaps the most important for the present volume. If, as we note, questions of science in place and moving over space have grown ever more important, it is also worth emphasizing that focusing on the binary of space and place means that other geographical imaginaries and tools of geographical categorization, most prominently scale, have been neglected and need to be brought into critical tension with the study of the history of science.

Scale and the Nineteenth-Century Sciences

The preeminence of "space" and "place" in the geographical study of science over the past several decades—and their terminological elasticity—has been the subject of critical review.[33] So, too, has scale, a term similarly central to the discipline of geography. The principal distinguishing feature of this theoretical and empirical reappraisal, which takes its inspiration from social constructivism and work in the "spatial turn" more widely, has been to reject notions of scale as a given "thing." Where scale once meant, in its simplest sense, a level of representation and, in that sense, primarily denoted levels of representation in terms of spatial extent, it is now more understood as relational, as something produced through social interaction. Thus, for geographer Sally Marsden,

> what is consistent about the recent interest in scale among social theorists in geography is the commitment to a *constructionist* framework and the rejection of scale as an ontologically given category. In these recent social theoretical studies, the fundamental point being made is that scale is not necessarily a preordained hierarchical framework for ordering the world—local, regional, national and global. It is instead a contingent outcome of the tensions that exist between structural forces and the practices of human agents. . . . Social theorists' attempts to address scale focus on understanding the processes that shape and constitute social practices at different levels of analysis.[34]

In Marsden's work and in that of other geographers, application of this revivified sense of scale and of scale-making as a relational process is directed primarily at explanation of the urban, political, and economic transformation of capitalist social relations, globally and in modern context. Thus, "scale construction is a political process endemic to capitalism," and "understanding the ways in which scales are constructed appears to be very much bound up with understanding the impacts and implications of globalization—which I take to mean the restructuring of world capitalism engendered by the economic crises of the 1970s."[35]

We would argue that this critical sense of scale is as applicable to explanation of science's geographies in historical context as it is to capitalism's geographies in the modern world. Our use of the terms "local," "national," and "global" to structure this collection reflects this view. What we intend by this threefold usage is not advocacy of strictly defined and separate hierarchical levels of spatial representation. We mean, rather, to highlight degrees of social and material interaction, as what passed for science at any one moment or venue was relationally determined, for example, by proximity and the particular characteristics of that place then (without ignoring the implications of social and geographical distance from others). We mean by it the nature of direct encounter (while recognizing others' mediated engagement with science through reading about it later and elsewhere). We mean by it the strength of connections between science's operatives in different places (as, for example, they worked in specific venues, such as a laboratory, on problems shared by others elsewhere). That is, what passed for science in its different locales—to take one of our three terms—was made by the nature and strength of the circumstances in that place (whether laboratory, home, or garden), and by the specific nature of the connections with related circumstances elsewhere. Notions of what the scientific locale was, or of the nation or the "national" or the "global" as frames of reference, are produced by how practitioners in the past worked and configured their work, and by how we, as modern researchers, configure that work as different objects of interest in different scalar and geographical imaginaries. Scales are not prefigured levels of spatial representation and so, by implication, have different epistemological significance. Thus understood, the distinction between "locality" and "globality" identified by Galison seems less problematic since, with scale and scale-making cast as relational, they are not predetermined analytic categories or "levels" for thinking about the geographies of science. Further, this means that some texts or practices of science may have a different explanatory role according to the scale at which the historian of science is operating. Scale, then, allows us to understand historical geographies of science but also to build different narratives about those practices.[36]

Several points follow from these remarks. With reference to the present volume, it is clear that chapters other than those by Mayhew and Sherratt and Finnegan speak to issues of "locale," just as those chapters have national and global resonance, albeit with less emphasis. Withers's attention to the meetings of the International Geographical Congress is, in one sense, a study in locale, of specific institutional venues in which participants' discursive conduct was regulated in order to achieve a given global outcome that would have—did have—different significance for different nations. Likewise, Rupke's chapter on structuralist evolution in Germany highlights the importance of the University of Jena, and of certain professors there, in giving institutional and individual succor to an ideology of politicized racism with global consequences here read as "national." The labels we use here as editors speak to our authors' intended emphasis—to their geohistoriographies—not a priori to defined and discrete levels of geographical analysis.

The critical attention to place, space, and scale has its counterpart in studies of "the national" within historical studies generally, not just the historical geographies of science. Galison's treatment of the "national" was implicit in his discussion of "locality" and "globality." Others have been more forthright. One historian of science has spoken of an apparent "end to national science," so prominent have been local studies in particular.[37] Are we witnessing the death of the national as a way of reading science geographically? Increased attention in historical studies to a variety of terms—"cosmopolitanism," "internationalism," "internationalization," and "transnational," as well as to "global history" and its variant "globality"—might suggest so. Attention in historical explanation, and in the historical study of science, to networks and flows, and to the mobility of ideas and materials between places and different social groups might suggest so, too.[38] At the same time, however, one effect of thinking above and beyond the nation, so to say, has been to produce the national even more acutely. The editors of a set of essays on transnationalism in the history of science note this paradox. Because transnational history shares with world history and global history the concern to abandon Euro- and US-centric viewpoints by focusing on how the circulation of people, objects, and ideas has differently cast nations' roles in the past, the effect has been to revivify the national: "transnational history pays more attention to nation states than do its alternatives."[39]

Exactly the same intellectual arc—prioritizing the global, transnational, and mobile only to find the national reasserting itself—has been described in intellectual history by efforts to forge a global intellectual history critical of a previous generation of contextual inquiry.[40] The chapters by Rupke, Noll, and Numbers can be read in this context. The "national" they separately invoke speaks not only to the boundaries of the nation-state in which their object of study is configured but also to the particular locale in and from which debates on racial difference or varieties of evolutionary theory came to have national, even transnational, significance and reach. Similarly, the chapter by Damodaran speaks to a particular extension of the nation—the colony—as a space in which ideas about a global "environmental anxiety" were articulated from the 1860s.

Rather than pronounce upon the elision or not of the "national," or evoke notions of "place" "space," and "locale" as straightforward analytic categories when they are not, we want, rather, to insist on the utility of these terms precisely because they provide an appropriate lexicon for thinking geographically about science and the complexities of the key terms "geography" and "science." To admit that these terms have fluidity of meaning is not to endorse imprecision. It is to insist on the value of the terms, to expect that authors make clear what they mean by them in their research, and, when appropriate, that the relational connections between the terms of space and place and the scales they signify are apparent. It is also to insist that we unpack the meaning of these key terms in historical and geographical context for past actors, scientific and otherwise. Commentators have observed the same of that other geographical term commonly employed in explanation of science's spatial dimensions, namely "islands." Islands are commonly sites where the global ebb and flow of capital and scientific knowledge has particular expression. In turn, islands' experiences have worldly consequences: "It is because islands occupy this conflicted relation that they became important sites for the creation of modern science as a universal fact and also as knowledge which can scale the local to the global." At the same time, and speaking of particular islands' geographies, they note how "it is also critical to deconstruct this view of the island, as laboratory, archive and field, and to place the islands of the Indo-Pacific and their various knowledge traditions in a much more variegated and plural history."[41]

Attention in these ways to science's placed and scaled geographies—whether on islands, in laboratories, or on board ships—is not just a matter of the production of science or of "professional" practitioners. It demands also that consideration be given to how the science in question was made mobile, how it managed to travel across space and between interested parties, and how it was received by particular audiences. Damodaran's essay is particularly illustrative in this respect. Attention to the "professional" sciences and their practice and performance as each became disciplined in the course of the nineteenth century should not preclude study of science's popularization among that equally important and fluid category, "the public." Here, too, much has been accomplished in the past thirty years or so.

In their important 1994 review of these issues, historians of science Roger Cooter and Stephen Pumfrey noted how terms such as "the public" and "audience" were commonly defined as consequences of a preexisting corpus of knowledge in historical (and geographical) context, whose consequences were measured against scientists' views rather than against audiences'.[42] Because conceptions of the "public understanding of science" tended to trade upon historical and sociological formulations of the "popular" as passive lay consumption of learned products, "science as product was boxed away from society, its production epistemologically privileged, its audience conceived as entirely yielding to new forms of natural knowledge."[43] For Cooter and Pumfrey, the history (and, we would argue, the geography) of popular science remained fragmented and in thrall to "authorized science," notwithstanding the local production of knowledge, differences across ideological divides, and the various discursive strategies implicated in the creation of public authority for scientific knowledge.

A range of work coincident with the spatial turn in science and that critical reevaluation of key geographical terms summarized earlier has extended understanding of the practices by which science was publicly received and made popular. For Jonathan Topham in his 2009 review of popular science, "science in popular culture" could not be made the subject of a separate subdiscipline, since to do so would involve naive disregard of the manner in which the "elitism of scientific discourse immediately delegitimizes popular experiences and epistemologies of nature."[44] But, Topham insists, reintegrating the histories (and geographies) of science's popularization and of science in popular culture within a reconceptualized history of science that is attentive to science's geographies is possible if we consider science, to use Secord's phrase, as "a form of communicative action."[45] For Topham, Secord's emphasis in his 2004 paper "Knowledge in Transit" had several effects:

> Secord's new approach thus places the practices of science popularization firmly within the process of knowledge making, alongside such other communicative practices as talking and note taking in laboratory or field, writing research papers, defending research findings within learned societies and congresses, advising on government policy as an expert witness, and teaching students in classrooms and laboratories. Similarly, it makes the place of science in popular culture central to understanding how the knowledge claims of scientific elites were established in relation to the full range of competing knowledge claims within a culture. In Secord's hands, the history of popular science disappears as a disciplinary subfield, only to reappear at the heart of the discipline.[46]

Others have likewise turned to the different forms taken in the transmission and reception of elite and non-elite science, and to science's "new audiences." Thus historian of the earth sciences Ralph O'Connor advocated attention to the literary genres of the acts of communication.[47] To genre we can add those questions of transmission embedded in issues relating to the translation of science.[48] For Bernadette Bensaude-Vincent, considerations of the mutual configuration of "science" and "public" may be enhanced by attention to "new reception history" (the phrase is O'Connor's) and, interestingly, by more local studies on the cultures of science—this to adopt one element of Galison's "problem" in contrast to Secord's view about a return to "grand narrative" in the history of science.[49] Secord's call is for a focus upon the practices and processes of science, less perhaps upon the definitive claims made by scientists. It is noteworthy in this respect that the essays in Aileen Fyfe and Bernard Lightman's 2007 *Science in the Marketplace* are structured around the headings of "Orality," "Print," and "Display."[50]

To assert with confidence that science studied spatially is now an established pursuit is not to prescribe what the future range and focus of such study should be. The chapters presented in the present volume build upon existing work and extend it with reference to gender, practice, race, politics, and methods of interpretation. The glossary of analytic terms employed generally includes but is not limited to notions of "place," "space," "scale," or, as here, "local," "national," and "global." Their application to specific settings, individual sciences, given personnel, or particular practices—reading, speaking, publishing, note-taking "in the field"—is always likely to be a matter of the researcher's preferences more than of any agreed-upon listing of key topics, some with presumed greater importance than others.

Reviewing in 2010 what he called "landscapes of knowledge"—and looking back on decades of achievement—David Livingstone sought to supplement the agenda for geographical studies of science with reference to "the role of landscape in knowledge enterprises, to the political ecology of science, to the critical significance of print culture in the circulation of scientific claims, and to the place—places—of speech in scientific culture."[51]

It is, of course, possible to add to this list, for the nineteenth century and for other periods. We need to know more about the role of gender in the venues and receptive spaces of nineteenth-century science: Is Wheeler-Cuffe's involvement in botany as much about class and status as it is obviously about gender and botany as a "safe" science for women? Did institutional practice or particular science's epistemic cultures vary in this respect? Did science in popular and public culture? How did indigenous and colonial knowledge spaces interact? Print culture and speech spaces were undoubtedly important, but so were instrumental cultures and the development and implementation of practices of measurement, instrumentation, and experimentation. Cultures of scientific representation—of mapping, statistical diagrams, photography—have an already established place and further rich potential. Are we in a position to write the geographies of individual sciences in national context in ways, à la Rupke and Numbers, that speak to locales within the nation and to their global implications over time? Colonial settings and concerns about climate and environment in the colonies have an evident historical geography in and for Britain: Is this true of different European empires? Should we turn our attention to technology, or, at least, to the practical implications intended by scientific progress and to their uneven geographies? The term "applied science" emerged in mid-nineteenth-century Britain, for example, to speak to particular narratives of progress, new educational institutions, and technically minded curricula and notions of individual achievement by "heroes of science" and "captains of industry."[52] Following Mayhew and Sherratt, how far would the published work of other leading scientists disclose the importance of locale if subject to a spatial hermeneutics? As Kennedy reminds us, reliable knowledge about the world in the nineteenth century was not won easily, nor once secured did it always safely travel. The study of science's failures—their why, when, and where—would illuminate science's social authority and spatial significance no less than its successes.

If this volume has brought up these sorts of queries and avenues for specific future research at the interface of spatiality and science in the nineteenth century, it also encourages us to speculate about the broader trends that future debate may exhibit. If this volume shows the fruitfulness of attending to scale in our histories of science, it is quite possible that other key conceptual categories from contemporary geography may be of use. For example, just as scale has been revived and made anew by geographers, so the "region" is no longer a term defined by interwar debates in Anglophone and European geography, with the past two or three decades seeing the flourishing of "new regional geographies" and new understandings of the region thanks to the likes of Doreen Massey.[53] A regional historical geography of science could take its cue from Massey, looking to the conjunction of modes of production and rounds of investment and disinvestment as building the tableaux in which science was distinctively produced, received, and disseminated at local scales. A new regionalism could push historians of science beyond seeing sites or locales as "black boxes," themselves incapable of being explained or analyzed, and toward an approach to small-scale histories of science that have greater explanatory purchase on the question of how and why sites and locales produce and receive knowledge (scientific and other) differently. Similar comments could be applied to other terms and metaphors of spatial analysis, notably "network"—which has only been received as a term of Latourian art thus far by the history of science—and "constellation." Above all, spatiality as a grounded reality and a metaphor has proved uniquely productive as an entrée to new approaches to the history of science over recent decades; there are many other spatial terms that could prove equally pregnant with meaning for the foreseeable future. If, as the philosopher of history Hans Blumenberg argued, metaphors are productive of meaning, the conjunction of spatiality and the history of science looks well set to continue to build such inquiries.[54]

Beyond spatial metaphors and concepts, geographical discourse could also productively add other sources of inspiration for new spatial histories of science. For example, the turn to "more than human" geographies that blur boundaries of animals and humans and of nature and culture could prove important in complicating the narratives we build about the history of science. If animals were key to scientific experiment and demonstration, they have been silenced as actors in the development of science, something a "more than human" history of science might complicate. If nature is a metaphor as redolent of the arrogation of scientific authority as Daston has chronicled, it can be used in reverse fashion to more emancipatory ends of questioning science's anthropocentrism and overturning the binaries on which that has been predicated.[55]

Above all, through both its conceptual lexicon and its debates about nature and culture, geography can continue to offer a prism through which historians of science can build new understandings of their subject matter. In this volume, we make good on this claim by attending to the role of scale as a category that helps us to understand the history of science and build new narratives about that history in dialogue with but also pushing beyond the conjunctions of science and spatiality that David Livingstone has pioneered.

Notes

1. Peter Galison, "Ten Problems in History and Philosophy of Science," *Isis* 99 (2004): 111–24, quote on 119. Galison's other problems were those of context (problem 1), purity and danger (2), historical argumentation (3), "fabricated fundamentals" (which together make up problems 4 and 5), political technologies (6), "relentless historicism" (9), and scientific doubt (10).

2. Galison, "Ten Problems," 122; emphasis original.

3. Barney Warf and Santa Arias, "Introduction: The Reinsertion of Space in the Humanities and Social Sciences," in *The Spatial Turn: Interdisciplinary Perspectives*, ed. Barney Warf and Santa Arias (London: Routledge, 2009), 1–10, quote on 1.

4. In rough chronological order, reviews of the spatial study of science would include Adir Ophir and Steven Shapin, "The Place of Knowledge: A Methodological Survey," *Science in Context* 4 (1991): 3–21; David Turnbull, "Reframing Science and Other Local Knowledge Traditions," *Futures* 29 (1997): 551–62; Crosbie Smith and Jon Agar, eds., *Making Space for Science: Territorial Themes in the Shaping of Knowledge* (Basingstoke: Macmillan, 1998); Steven Harris, "Long-Distance Corporations, Big Sciences, and the Geography of Knowledge," *Configurations* 6 (1998): 269–305; Steven Shapin, "Placing the View from Nowhere: Historical and Sociological Problems in the Location of Science," *Transactions of the Institute of British Geographers* 23 (1998): 5–12; Jan Golinski, *Making Natural Knowledge: Constructivism and the History of Science* (Cambridge: Cambridge University Press, 1998); Peter Meusburger, "The Spatial Concentration of Knowledge: Some Theoretical Considerations," *Erdkunde* 54 (2000): 352–64; Charles W. J. Withers, "The Geography of Scientific Knowledge," in *Göttingen and the Development of the Natural Sciences*, ed. Nicolaas Rupke (Göttingen: Wallstein Verlag, 2002), 9–18; David N. Livingstone, *Putting Science in Its Place: Geographies of Scientific Knowledge* (Chicago: University of Chicago Press, 2003); James A. Secord, "Knowledge in Transit," *Isis* 95 (2004): 654–72; Simon Naylor, "Introduction: Historical Geographies of Science—Places, Contexts, Cartographies," *British Journal for the History of Science* 38 (2005): 1–12; Richard Powell, "Geographies of Science: Histories, Localities, Practices, Futures," *Progress in Human Geography* 31 (2007): 309–30; Diarmid A. Finnegan, "The Spatial Turn: Geographical Approaches in the History of Science," *Journal of the History of Biology* 41 (2008): 369–88; David N. Livingstone, "Landscapes of Knowledge," in *Geographies of Science*, ed. Peter Meusburger, David N. Livingstone, and Heike Jöns (Dordrecht: Springer, 2010), 3–22; Charles W. J. Withers and David N. Livingstone, "Thinking Geographically about Nineteenth-Century Science," in *Geographies of Nineteenth-Century Science*, ed. David N. Livingstone and Charles W. J. Withers (Chicago: University of Chicago Press, 2011), 1–19.

5. Although they are not by intention reviews of this field, several volumes in the Knowledge and Space series, edited by Peter Meusburger and colleagues and published by Springer on behalf of the Klaus Tschira Foundation, illustrate well the geographical study of science in particular, and of knowledge in general; see especially volumes 3, 7, 9, 10, and 11 in this series: Peter Meusburger, David N. Livingstone, and Heike Jöns, eds., *Geographies of Science* (Dordrecht: Springer, 2010); Peter Meusburger, Derek Gregory, and Laura Suarsana, eds., *Geographies of Knowledge and Power* (Dordrecht: Springer, 2015); Peter Meusburger, Benno Werlen, and Laura Suarsana, eds., *Knowledge and Action* (Dordrecht: Springer, 2017); Heike Jons, Peter Meusburger, and Michael Heffernan, eds., *Mobilities of Knowledge* (Dordrecht: Springer, 2017); and Johannes Glückler, Emmanuel Lazega, and Ingmar Hammer, eds., *Knowledge and Networks* (Dordrecht: Springer, 2017).

6. For early declarations, see, for example, David N. Livingstone, "The History of Science and the History of Geography: Interactions and Implications," *History of Science* 22 (1984): 271–302.

7. David N. Livingstone, "Science and Religion: Foreword to the Historical Geography of an Encounter," *Journal of Historical Geography* 20 (1994): 367–83, quote on 367; emphasis original.

8. David N. Livingstone, "The Spaces of Knowledge: Contributions towards a Historical Geography of Science," *Environment and Planning D: Society and Space* 13 (1995): 5–34.

9. Livingstone, *Putting Science in Its Place*; David N. Livingstone, "Science, Text and Space: Thoughts on the Geography of Reading," *Transactions of the Institute of British Geographers* 35 (2005): 391–401; David N. Livingstone, "Science, Site and Speech: Scientific Knowledge and the Spaces of Rhetoric," *History of the Human Sciences* 20 (2007): 71–98; David N. Livingstone, *Adam's Ancestors: Race, Religion, and the Politics of Human Origins* (Baltimore: Johns Hopkins University Press, 2008); David N. Livingstone, *Dealing with Darwin: Place, Politics, and Rhetoric in Religious Engagements with Evolution* (Baltimore: Johns Hopkins University Press, 2014); David N. Livingstone, "Debating Darwin at the Cape," *Journal of Historical Geography* 52 (2016): 1–15.

10. Bernard Lightman, ed., *Victorian Science in Context* (Chicago: University of Chicago Press, 1997); David Cahan, ed., *From Natural Philosophy to the Sciences: Writing the History of Nineteenth-Century Science* (Chicago: University of Chicago Press, 2003); James A. Secord, *Victorian Sensation: The Extraordinary Publication, Reception, and Secret Authorship of Vestiges of the Natural History of Creation* (Chicago: University of Chicago Press, 2000); Livingstone and Withers, eds., *Geographies of Nineteenth-Century Science.*

11. Deborah R. Coen, *Climate in Motion: Science, Empire, and the Problem of Scale* (Chicago: University of Chicago Press, 2018); Nathan F. Sayre, *The Politics of Scale: A History of Rangeland Science* (Chicago: University of Chicago Press, 2017); and Lydia Barnett, *After the Flood: Imagining the Global Environment in Early Modern Europe* (Baltimore: Johns Hopkins University Press, 2019).

12. This phase is taken from David N. Livingstone's *Putting Science in Its Place* (183; emphasis original) as part of his injunction that biography, a staple if much-changed feature in the history of science, should treat a subject's life *geography*, that is, elucidate its placed dimensions, rather than that more traditional focus of a life disclosed over time.

13. On the term "reputational geographies," see David N. Livingstone, "Politics, Culture, and Human Origins: Geographies of Reading and Reputation in Nineteenth-Century Science," in Livingstone and Withers, *Geographies of Nineteenth-Century Science*, 178–202.

14. The distinction between "laboratory" and "field" is a much-vaunted but, we suggest, problematic distinction within work in the geographies of science since it tends to separate two "categories" of space and not address connections between them. For excellent discussions of the issue, see, for example, Finnegan, "Spatial Turn"; and Powell, "Geographies of Science"; and, at greater length, Golinski, *Making Natural Knowledge*. For a fuller-length analysis of the issue, see the essays in Henrika Kuklick and Robert E. Kohler, eds., *Science in the Field, Osiris* 11 (1996), and, with particular reference to biology, Robert E. Kohler, *Landscapes and Labscapes: Exploring the Lab-Field Border in Biology* (Chicago: University of Chicago Press, 2002). For an example of nineteenth-century science (genetics) in a domestic context, see Donald L. Opitz,

"Cultivating Genetics in the Country: Whittinghame Lodge, Cambridge," in Livingstone and Withers, *Geographies of Nineteenth-Century Science*, 73–98.

15. On the example of meteorology, see (in a wide literature on this subject), Katherine Anderson, *Predicting the Weather: Victorians and the Science of Meteorology* (Chicago: University of Chicago Press, 2005); Paul Edwards, "Meteorology as Infrastructural Globalism," *Osiris* 21 (2006): 229–50; and Martin Mahony, "For an Empire for 'All Types of Climate': Meteorology as an Imperial Issue," *Journal of Historical Geography* 51 (2016): 29–39. Geographer Simon Naylor has paid especial attention to the different geographies of meteorology in the nineteenth century: see Simon Naylor, "Nationalizing Provincial Weather: Meteorology in Nineteenth-Century Cornwall," *British Journal for the History of Science* 39 (2006): 407–33; Simon Naylor, "Log Books and the Law of Storms: Maritime Meteorology and the British Admiralty in the Nineteenth Century," *Isis* 106 (2015): 771–97; and Simon Naylor, "Thermometer Screens and the Geography of Uniformity in Nineteenth-Century Meteorology," *Notes and Records of the Royal Society of London* 73 (2018): 203–221.

16. On historical geographies of science in a global context, see the essays in Diarmid A. Finnegan and Jonathan J. Wright, eds., *Spaces of Global Knowledge: Exhibition, Encounter and Exchange in an Age of Empire* (Farnham: Ashgate, 2015). On science in its Victorian context in which many of the studies are of particular practices with assessment of their geographical reach, see the essays in Bernard Lightman, ed., *Victorian Science in Context* (Chicago: University of Chicago Press, 1997).

17. On expeditionary science—and, indeed, how expeditionary field sites and practices might also be construed as laboratories *en plein air*—see the essays in Kristian H. Nielsen, Michael Harbsmeier, and Christopher J. Ries, eds., *Scientists and Scholars in the Field: Studies in the History of Fieldwork and Expeditions* (Aarhus: Aarhus University Press, 2012). For illustrations of expeditionary science and its counterpart, "critical" or "armchair geography," which was based on the largely sedentary examination of textual sources rather than firsthand observation, see Lawrence Dritsas, "Expeditionary Science: Conflicts of Method in Mid-Nineteenth-Century Geographical Discovery," in Livingstone and Withers, *Geographies of Nineteenth-Century Science*, 255–77; Charles W. J. Withers, "Mapping the Niger, 1798–1832: Trust, Testimony and 'Ocular Demonstration' in the Late Enlightenment," *Imago Mundi* 56 (2004): 170–193; and David Lambert, *Mastering the Niger: James Macqueen's African Geography and the Struggle over Slavery* (Chicago: University of Chicago Press, 2013).

18. On this, see the essays in Dane Kennedy, ed., *Reinterpreting Exploration: The West in the World* (Oxford: Oxford University Press, 2014).

19. David Cahan, "Looking at Nineteenth-Century Science: An Introduction," in *From Natural Philosophy to the Sciences*, 3–15, quote on 3.

20. Cahan, "Looking at Nineteenth-Century Science," 4.

21. See Theodor M. Porter and Dorothy Ross, eds., *The Cambridge History of Science: Volume 7: The Modern Social Sciences* (Cambridge: Cambridge University Press, 2003); Johan Heilbron, *The Rise of Social Theory* (Cambridge: Polity Press, 1995); Johan Heilbron, Lars Magnusson, and Björn Wittrock, eds., *The Rise of the Social Sciences and the Formation of Modernity: Conceptual Change in Context, 1750–1850* (Dordrecht: Kluwer Academic Publishers, 1998); and Lawrence Goldman, *Science, Reform and*

Politics in Victorian Britain: The Social Science Association, 1857–1886 (Cambridge: Cambridge University Press, 2002).

22. John Merz, *A History of European Thought in the Nineteenth Century* (Edinburgh: William Blackwood and Songs, 1904–12); John Bernal, *Science in History* (London: Faber and Faber, 1954); Joseph Ben-David, *The Scientist's Role in Society: A Comparative Study* (Englewood Cliffs, NJ: Prentice-Hall, 1971).

23. Cahan, "Looking at Nineteenth-Century Science," 11.

24. On these issues, see J. B. Morrell, "Professionalisation," in *Companion to the History of Modern Science*, ed. Roger C. Olby, Geoffrey N. Cantor, John R. R. Christie, and Martin J. S. Hodge (London: Routledge, 1990), 980–89. In arguing for the emergence of science as "a regular vocational pursuit," Morrell advanced six features. The first was an increase in the number of full-time paid positions. Second, specialist qualifications were established. These functioned as the public certification of scientific competence. Third, training procedures were developed, especially in and through the university laboratory. The fourth feature was the rapid growth of specialization in published research, evident in "the development of esoteric technical languages, in the greater application of analytical mathematics and in the deployment of arcane experimental techniques. . . . In other words, various sciences were being demarcated as specific areas of skill, knowledge and expertise" (983). Fifthly, there was growing group solidarity and self-consciousness among science's practitioners. The sixth feature was the development of various reward systems geared to recognize best practice and peer esteem: the conferment of honors, fellowship of a scientific society, and so on.

25. Cahan, "Looking at Nineteenth-Century Science," 14.

26. On the use of this term with reference to attempts to spatialize history, see Robert Mayhew, "Historical Geography 2009–2010: Geohistoriography, the Forgotten Braudel and the Place of Nominalism," *Progress in Historical Geography* 35 (2010): 409–21. See also Paul Ethrington, "Placing the Past: 'Groundwork' for a Spatial Theory of History," *Rethinking History* 11 (2007): 465–93, and Paul Stock, "History and the Uses of Space," in *The Uses of Space in Early Modern History* (London: Palgrave, 2015), 1–18. For further discussions of the challenges posed by historical explanation in science studies and, in part, the spatialization of science as a historical project, see the essays in Robert E. Kohler and Kathryn M. Olesko, eds., *Clio Meets Science: The Challenges of History*, *Osiris* 27 (Chicago: University of Chicago Press, 2012).

27. We take these points from Nils Güttler, *Das Kosmokop: Karten und ihre Benutzer in der Pflanzengeographie des 19 Jahrhunderts* (Göttingen: Wallstein Verlag, 2014).

28. Joseph D. Hooker, "On Geographical Distribution," in *Report of the Fifty-First Meeting of the British Association for the Advancement of Science* (London: John Murray, 1882), 727–38, quotes on 731 and 733.

29. We take these points from Nicolaas A. Rupke, "Humboldtian Distribution Maps: The Spatial Ordering of Scientific Knowledge," in *The Structure of Knowledge: Classifications of Science and Learning since the Renaissance*, ed. Tore Frängsmyr (Berkeley: University of California Press, 2001), 93–116. These points about botany's placed making are not to overlook those other places in which that science was made: either in, say, the "microhistorical" practices and spaces of books of plants where pressed specimens were used instructively in teaching the subject or, in an institutional and

biographical sense, in the spaces of botanical gardens. On the first, see Anne Secord, "Pressed into Service: Specimens, Space, and Seeing in Botanical Practice," in Livingstone and Withers, *Geographies of Nineteenth-Century Science*, 283–310. On the second, see Jim Endersby, *Imperial Nature: Joseph Hooker and the Practices of Victorian Science* (Chicago: University of Chicago Press, 2008).

30. Simon Winchester, *The Map That Changed the World: A Tale of Rocks, Ruin and Redemption* (London: Viking, 2001); Martin J. S. Rudwick, *Bursting the Limits of Time: The Reconstruction of Geohistory in the Age of Revolution* (Chicago: University of Chicago Press, 2005).

31. Withers and Livingstone, "Thinking Geographically about Nineteenth-Century Science," in *Geographies of Nineteenth-Century Science*, 2; emphasis original.

32. These points are taken from, and more fully examined in, Robert Mayhew, "A Tale of Three Scales: Ways of Malthusian Worldmaking," in Stock, *The Uses of Space in Early Modern History*, 197–226, esp. 197–201.

33. For a discussion and review of these terms, see Charles W. J. Withers, "Place and the 'Spatial Turn' in Geography and History," *Journal of the History of Ideas* 70 (2009): 637–58. See also Finnegan, "Spatial Turn."

34. Sallie A. Marston, "The Social Construction of Scale," *Progress in Human Geography* 24 (2000): 219–42, quote on 220.

35. Marston, "Social Construction of Scale," 221.

36. Mayhew, "Tale of Three Scales."

37. Lewis Pyenson, "An End to National Science: The Meaning and Extension of Local Knowledge," *History of Science* 40 (2002): 251–90.

38. On this point, see the emphasis placed upon science's making as a form of "communicative action" in Secord, "Knowledge in Transit."

39. Simone Turchetti, Nestór Hernan, and Soraya Boudia, "Introduction: Have We Ever Been 'Transnational'? Towards a History of Science across and beyond Borders," *British Journal for the History of Science* 45 (2012): 319–26, quote on 322.

40. See Samuel Moyn and Andrew Sartori, eds., *Global Intellectual History* (New York: Columbia University Press, 2013); Willibald Steinmetz, Michael Freeden, and Javier Fernandez Sebastian, eds., *Conceptual History in the European Space* (New York: Berghahn, 2017); Rosario Lopez, "The Quest for the Global: Remapping Intellectual History," *History of European Ideas* 42 (2016): 155–60; and J. G. A. Pocock, "On the Unglobality of Contexts: Cambridge Methods and the History of Political Thought," *Global Intellectual History*, January 23, 2019 https://doi.org/10.1018/23801883.2018.1523997.

41. Pablo F. Gómez and Sujit Sivasundaram, "Epilogue," *British Journal for the History of Science* 51 (2018): 679–86, quote on 679. Their paper is the last in a special issue devoted to "Science and Islands in Indo-Pacific Worlds." On this point about islands as laboratories, see Richard Grove, *Green Imperialism: Colonial Expansion, Tropical Island Edens and the Origins of Environmentalism, 1600–1860* (Cambridge: Cambridge University Press, 1995), and Beth Greenhough, "Imagining an Island Laboratory: Representing the Field in Geography and Science Studies," *Transactions of the Institute of British Geographers* 31 (2006): 224–37. For one particularly elegant illustration of the ways in which islands as "locale" were molded by global connections (and vice versa), see Sujit Sivasunaram, *Islanded: Britain, Sri Lanka, and the Bounds of*

an Indian Ocean Colony (Chicago: University of Chicago Press, 2013). For a wide-ranging survey of the epistemic value of islands, see Marc Shell, *Islandology: Geography, Rhetoric, Politics* (Stanford: Stanford University Press, 2014).

42. Roger Cooter and Stephen Pumfrey, "Separate Spheres and Public Places: Reflections on the History of Scientific Popularization and Science in Popular Culture," *History of Science* 32 (1994): 237–67.

43. Cooter and Pumfrey, "Separate Spheres and Public Places," 240.

44. Jonathan R. Topham, introduction to "Focus: Historicizing 'Popular Science,'" *Isis* 100 (2009): 310–18, quote on 311.

45. Secord, "Knowledge in Transit."

46. Topham, "Introduction," 311.

47. Ralph O'Connor, "Reflections on Popular Science in Britain: Genres, Categories, and Historians," *Isis* 100 (2009): 333–45, quote on 334–35.

48. Scott Montgomery, *Science in Translation: Movements of Knowledge through Cultures and Time* (Chicago: University of Chicago Press, 2000).

49. Bernadette Bensaude-Vincent, "A Historical Perspective on Science and Its 'Others,'" *Isis* 100 (2009): 359–68.

50. Aileen Fyfe and Bernard Lightman, eds., *Science in the Marketplace: Nineteenth-Century Sites and Experiences* (Chicago: University of Chicago Press, 2007). See also on this point, Bernard Lightman, *Victorian Popularizers of Science: Designing Nature for New Audiences* (Chicago: University of Chicago Press, 2007). On the importance of books and of reading cultures for the early Victorian period in Britain (and so not straightforwardly applicable to the rest of the nineteenth century or to non-British contexts), see James A. Secord, *Visions of Science: Books and Readers at the Dawn of the Victorian Age* (Chicago: University of Chicago Press, 2014).

51. Livingstone, "Landscapes of Knowledge," 18. Although Livingstone's essay is, in part, a review of work in the geographies of science, it is distinct in confidence and tone from his earlier declarations in this respect: see Livingstone, "The History of Science and the History of Geography," and Livingstone, "Science and Religion."

52. Robert Bud, "'Applied Science' in Nineteenth-Century Britain: Public Discourse and the Creation of Meaning, 1817–1876," *History and Technology* 30 (2014): 3–36.

53. Doreen Massey, *Spatial Divisions of Labour: Social Structures and the Geography of Production* (London: Hutchinson, 1984).

54. Hans Blumenberg, *Paradigms for a Metaphorology* (Ithaca, NY: Cornell University Press, 2010).

55. Lorraine Daston, *Against Nature* (Cambridge, MA: MIT Press, 2019).

PART ONE

LOCALE STUDIES

CHAPTER ONE

Locating Malthus's Essay

Localism and the Construction of Social Science, 1798–1826

ROBERT J. MAYHEW AND YVONNE SHERRATT

"Can the location of scientific endeavour make any difference to the conduct of science? And even more important, can it affect the content of science? In my view the answer to these questions is yes."[1] By the time that David Livingstone opened his influential book *Putting Science in Its Place* (2003) with these words, a generation of scholars in the history of science had already been mooting the same points and reaching the same conclusions. Indeed, Livingstone's book amounted to the distillation of a simple yet profound claim that the group had built: geographical matters of location, space, and scale are central to the creation, circulation, and reception of (scientific) knowledge. Scholarship in the history of science and of the disciplines more generally had long since deployed ideas such as national cultures of scholarship, regional or provincial schools of inquiry, and colonial science to inform their inquiry.[2] But it was only in the 1980s and 1990s that scholars such as David Turnbull and Steven Shapin began to generalize the point implicit in these diverse inquiries: *where* things happen historically speaking matters to *what* those said things are.[3] A set of important texts regarding the locatedness of science and scholarship in the Enlightenment, in the various European revolutions, political and intellectual, and in the nineteenth century consolidated this theoretical insight. At the same time, it added immeasurable depth to our understanding of the capacity this insight had to enrich our historical understanding of these areas of inquiry.[4] In the decade and a half since Livingstone affirmed the significance of spatiality to science, numerous further historical and empirical inquiries have followed confirming that position.[5] Furthermore, this insight has fanned out from the history of science to broader swaths of intellectual, social, and cultural history such that one can now see, for example, intense debates about the relative merits of imagining a singular "Enlightenment" or geographically variegated "Enlightenments."[6] If Peter Galison's claim that "the turn toward local explanation in the historical, sociological, and philosophical understanding of science may well be the single most important change in the last thirty years" seemed plausible in 2008, it seems all but self-evident a decade later.[7]

The present chapter seeks to add new dimensions to the burgeoning conjunction between spatiality and science by attending to Thomas Robert Malthus's celebrated *Essay on the Principle of Population*. There can be no disputing the *Essay*'s status as the most influential demographic treatise of all time. The notoriety of Malthus's work has guaranteed for its author a place in narratives about the history of economics and in the social sciences. The present inquiry seeks to investigate the extent to which new insights into Malthus's much-studied achievement in particular and the construction of social science more generally can be generated by accepting Livingstone's central contention that the location of intellectual activity cannot be divorced from its content. The question being asked is the extent to which the local contexts in which Malthus lived—his spatial biography, as it were—influenced the construction of his argument both as it was first conceived in the lively form of an enlightened polemic in 1798 and as it was then recast into a far longer and more empirically dense statement in the so-called Great Quarto edition of 1803, that edition being the core text around which subsequent iterations were produced until a final, fifth version in 1826, that being the last published in Malthus's lifetime. As this suggests, our central concern is not with the spatial *contexts* in which Malthus wrote but instead with the spatiality disclosed in the *texts* that Malthus produced. The questions being opened up here, then, are first, to what extent the places in which Malthus lived and traveled are disclosed by the various versions of the *Essay* he produced between 1798 and 1826, and, second, to what extent the modes of his argumentation reflected the local situations he experienced and observed.

Attending to these questions with regard to Malthus's *Essay* allows this inquiry to address four angles on the issue of the geography of science that have been relatively neglected by the burgeoning literature in the field. First, there is the local or locale, the place in which a scientist or scholar lives. While scholars have attended to the small-scale sites in which science is conducted—the laboratory, the field-station, and so forth—the places in which people actually live have been less scrutinized as potentially helping to understand the ideas they produce.[8] And yet, to the extent that empirical observation of the world is an essential platform for certain varieties of scholarly inquiry, surely it is worth contemplating where scientists and scholars lived and the extent to which this may have provided material that became embedded in their arguments. In turning to Malthus specifically, then, the question would be whether the places in which he lived and to which he had traveled make a clear contribution to the argumentation he deploys in the *Essay*.

Second, refocusing from where scientists and scholars worked to a broader question about where they lived and how this impacted their work is to take seriously an inquiry whose potential fruitfulness Livingstone flagged toward the close of *Putting Science in Its Place* in his call to show "greater sensitivity to the *spaces of a life*."[9] Livingstone suggested that the locational awareness of a geography of science needed to be complemented by attention to the spatial navigation inherent in any scholar's biography with a view to seeing how this life course—spatial as well as personal—was reflected in their achievements. Of course, and as Livingstone's comments at this point made clear, some work had already been done in this vein, notably Martin Rudwick's analysis of the Devonian controversy in geology and Desmond's analysis of the London circles in which the young Darwin moved and built his evolutionary ideas.[10] In the case of Malthus, we already have an authoritative biography by Patricia James to which only a few major additions have been made by archival discoveries over the past four decades.[11] But the extent to which the spatial aspect of Malthus's life course may have had an impact on his ideas, let alone their textual expression, has been entirely neglected with the exception of one powerful and important insight offered by James herself, which is discussed in this chapter. As such, despite Livingstone's call a decade ago to attend to spaces of life and despite a clear understanding of Malthus's life course, there still remains much work to be done to understand how the spaces of Malthus's life over the thirty years from his emergence on the public stage as an author in 1798 until the final iteration of his *Essay* in 1826 are imbricated in the texture of his arguments.

Thirdly, another avenue of inquiry David Livingstone proposed was a "geography of reading," "arguing for the fundamental importance of the spaces where reading literally *takes place*, for the knowledge that is produced in moments of textual encounter."[12] Livingstone's point was about the intersection of geography—where reading takes place—with reception studies as pioneered by Wolfgang Iser and Hans Robert Jauss. In this vein, he has in recent work traced in the reception of Darwin and the ways in which the thesis of the *Origin of Species* was understood differently according to the various intellectual discourses in whose context it was inserted in different places.[13]

And yet, there is a broader intersection between the geography of science as an idea and textual interpretation and reception that has not received any attention. As well as attending to how different individuals in different geographical-cum-intellectual contexts receive a text, we as scholars can also use the core idea that location affects the construction of a scholarly argument to build what might be called a *spatial hermeneutic*. By this we mean that a reading strategy, a hermeneutic angle by which to approach a text, can be built attending to how any given text reveals the spaces of life in which an author has lived and through which they have traveled. This spatial hermeneutic is exactly the tack that the present inquiry takes, using a spatial reading strategy to find different webs of meaning and signification in Malthus's *Essay*.

Fourth and finally, in attending to Malthus's *Essay*, the present inquiry extends the remit of the geography of science approach to the social sciences. In general, the geographical analysis of the production, circulation, and reception of knowledge has emerged from and focused on the analysis of the history of the physical, natural, and experimental sciences. As stated earlier, some of the insights of a spatially sensitive historiography have been carried over into debates in intellectual history, but historians of the social sciences have been comparatively reticent in taking up this approach. As evidence of this point, if one looks at the only other recent collection of essays bringing together geographical sensitivity and nineteenth-century knowledge claims beyond the present volume, only one of its fifteen substantive chapters clearly takes on a topic from the social sciences, that being Livingstone's chapter on the spatial variation in debates about human origins, treating as it does the intellectual territory of social Darwinism and anthropology. Other chapters address elements of social scientific concern via the rather eclectic British Association for the Advancement of Science and questions of museum curation and presentation. Overall, the point remains: social science is comparatively marginal to the analysis of the geographies of knowledge, this despite the nineteenth century in most historiographical readings seeing the birth of the social sciences.[14] As such, rereading Malthus's *Essay* in light of the concerns and methods emerging from geographical readings of the history of science both analyzes the applicability of these ideas in a different intellectual context and hopefully offers new readings of the contested historiographical topic of the "birth of the social sciences" and Malthus's place in that nativity, a topic to which we turn next.

Historiography, Rhetoric, and Malthus's Place in the "Birth" of Social Science

A conventional historiography of the social sciences places its mature emergence in the mid- to late nineteenth century, evident in the institutional emergence of an educative apparatus at European and North American universities and through the combined intellectual achievements of the triumvirate of Karl Marx, Max Weber, and Émile Durkheim, who are posited as building a "classical" sociology and social theory.[15] Yet, for many decades now, historians of the social sciences have also been keen to point to longer historical genealogies for their subject matter. Simplifying, it can be said that they have detected three separate temporal scales on which to chart the putative "birth" of the social sciences. First, and on the longest timescale, it has been pointed out by several scholars that inquiries into politics and governance, social organization, wealth, and structure are as old as the ancient Greeks and that we can detect the beginnings of more structured discourse about society from the emergence of "reason of state" arguments and political arithmetic in the sixteenth and seventeenth centuries.[16] Historians who take this approach over the longest timescale also tend to focus on the eighteenth century as seeing a discernible movement toward "modern" social thought through the emergence of a far more vital tradition of political economy and particularly through the development of so-called conjectural histories of social development among the literati of the French and Scottish Enlightenments.[17]

The locus classicus for this structure of argument was Meek's hugely influential study, *Social Science and the Ignoble Savage* (1976), which concluded that conjectural history was "clearly recognisable as an early prototype of the view which many [social] scientists would wish to put forward today."[18] On a somewhat shorter but overlapping timescale, more recent arguments about the birth of social science have attended to the two or so decades on either side of 1800 as seeing an epistemic break wherein the social sciences emerged. This argument, of course, has been driven by an engagement with Foucault's argument in *The Order of Things* but also through Reinhart Koselleck's idea of the era as a period of accelerated social, economic, and intellectual transformation, a *Sattelzeit*. On this basis, one collection of essays on the emergence of the social sciences argued that "there are reasons for carefully examining the ways in which the distinctively modern key concepts for the understanding of society emerged during the great transition in the late eighteenth and early nineteenth centuries."[19] Finally, drawing likewise on Foucault and Koselleck but looking on a shorter but still overlapping timescale, there has been an argument to pinpoint the emergence of modern social science to a yet more focused historical moment and geographical location: the maelstrom of argumentation sparked in France in the immediate aftermath of the Revolution. Thus both Wokler and Baker note that the term "social science" first emerged in the circle of scholars dubbed the Society of 1789 and arrayed around Condorcet: "The first printed use of the term *science sociale* in the year 1789 . . . indicated that a recognizably modern conception of the nature of the social sciences was developed in the course of the French Revolution."[20] Even acknowledging that the semantic usage of the term in 1789 did not match closely with modern conceptions of social science, the very emergence of the term has seemed indicative of a changing constellation of inquiries.[21] Wokler has gone further, following Foucault in seeing the term "*science sociale*" as undergoing "the epistemic break of metamorphosis . . . precisely in 1795" with its placement in the revolutionary *Institut national des Sciences et des Arts*, concluding that from this time "social science in particular came to acquire the meanings now associated with it as the central science of modernity."[22]

Chronological debates about *when* a modern social science emerged notwithstanding, there seems to be a shared understanding in this historiography about *what* distinguishes social science from heterogeneous inquiries into the social: the drive to find lawlike generalities about society. As such, historians of all stripes have sought the emergence of what might be called a rhetoric of universalism as indicative of social science. Thus conjectural history sought not just examples of societies moving through stages from hunter-gathering toward modern commercial activities but to claim that this was a trajectory all societies would follow. Furthermore, Enlightenment scholars sought to find the shared motivations behind social action and change, "self-interest," for example, being "regarded as a law of the social world akin to the principle of universal gravitation in the physical world." Generalizing from this, "Enlightenment students of society sought not just patterns but veritable laws describing underlying regularities explaining human behaviour in societies."[23] The model that literati on both sides of the English Channel sought to emulate in the study of society was that which Isaac Newton or his reception had created for the study of natural philosophy: "For thinkers of the Enlightenment . . . Newton's achievement was both a model and a challenge. He had shown what could be done and how to do it. The challenge was to emulate his work: to achieve for the moral or social sciences what he had done for natural science."[24] Whether the birth of social science is seen as a revolutionary flash or a longer burn of Enlightenment, its emergence is captured through uncovering the Newtonian trace of a universal inquiry into the social whose statements are designed to transcend locality, the specificity of the time and place in which they are produced.

Thomas Robert Malthus (1766–1834) has always fitted comfortably into this set of historiographical presuppositions about when social science emerged and what defines its modernity. Malthus slots neatly into the view that locates social science as emerging from the conjectural histories of the stages of social development, for example. All the editions of his *Essay* drew on this framework to contend for the ubiquity of his principle of population, this being delivered in four pithy chapters in the 1798 edition (chapters 3–6) and in two massive books in the Great Quarto edition of 1803 (books 1 and 2).[25] Likewise, Malthus's writing career spans from 1798 until his death and thereby stretches over much of the period identified as paradigmatic by scholars influenced by Foucault and Koselleck.[26] Finally, Malthus's *Essay* was, of course, originally couched as a riposte to the arguments of Condorcet, whom Baker and Wokler identify as key to the Francophone and revolutionary reading of the emergence of social science. On such a reading, even if Malthus stood at a very different point on the political spectrum, by using the very same tools and concepts of social analysis as Condorcet while turning them against him, Malthus fueled the revolutionary emergence of social science in this turbulent decade. For all three conceptualizations of the emergence of social science, then, "Malthus is a useful figure to mark the transition."[27]

More importantly, Malthus's modes of argumentation in the *Essay* are exemplary of the Newtonian drive to the discovery of universal laws about the social world that is deemed conceptually definitive of the project of social science. Malthus's aim, as is well known, was to find a set of invariant laws relating population to food availability and to argue that those laws would help to explain key phenomena in the social realm, notably class stratification, patterns of food- and resource prices, and labor value. In this inquiry, Malthus explicitly used a language of social mathematics, as had Condorcet, setting his thesis up in 1798 by proposing postulates of a lawlike character and rates of population and food increase of a fixed, quantitative nature (chapter 1). Behind this mathematical apparatus, as Malthus makes clear at several points in the *Essay*, is his reverence for Newton as a model for social inquiry.

In rebutting Condorcet's idea that the mere lack of evidence for the organic perfectibility of man does not disprove its possibility, for example, Malthus waxed indignant: "If this be the case, there is at once an end of all human science. . . . The grand and consistent theory of Newton, will be placed upon the same footing as the wild and eccentric hypotheses of Descartes. . . . The constancy of the laws of nature, and of effects and causes, is the foundation of all human knowledge."[28] Malthus, then, both through when he lived and how he expressed his ideas, has been a paradigmatic exemplar for a conventional framing of social science precisely in terms of its universalizing intellectual ambition and its effacing of the local and the geographically specific.

One of the key achievements of the geographical approach to the history of science canvassed previously has been to decenter and question the rhetoric of universal, disembodied, objective knowledge on which traditional discourses in the natural and physical sciences have traded. By emphasizing the geographical specificity of the production, circulation, and reception of scientific knowledge, that knowledge has inevitably been grounded, been brought back to the places and the personalities whence it sprung. One aim of the present chapter is precisely to offer a similar grounding to ideas in the social sciences, cutting against the universalizing presumptions of historiography addressing the birth of the social sciences. Just as from a postcolonial perspective Chakrabarty has sought to "provincialise" European modes of writing history, so from a spatial perspective we seek to show the local conditions within Malthus's universalizing discourse of Newtonian social laws, to show the moments in which Malthus's biographical and personal position—geographical and social—emerge in his text.[29]

While there has been surprisingly little such work on the history of the social sciences, one can point to several elements in recent work about Malthus's achievement that have begun to move in this direction. First, there is a brief but important aside in Patricia James's monumental biography. James notes Malthus's comment in the 1798 *Essay* that "the sons and daughters of peasants will not be found such rosy cherubs in real life, as they are described to be in romances. It cannot fail to be remarked by those who live much in the country, that the sons of labourers are very apt to be stunted in their growth" (1798, ch. 5: 29–30). As she commented, Malthus penned the *Essay* while he was living with his parents and his income came from being curate of the parish of Okewood, a remote place of extreme poverty: "When Malthus left Albury [the location of his parent's house] for the nine-mile ride to Okewood, along narrow bridle paths, he entered a world as remote from books as it was from turbot."[30] James opens the possibility here of reading the *Essay* not for universal laws of social science but for windows onto the author's own life experiences in 1798, located in Surrey. While this may seem to be the traditional fare of biography, it echoes back to Livingstone's point about attending to the spaces of life: where Malthus wrote the *Essay* was used to understand how he constructed his argument. This insight was taken further in Mayhew's *Malthus: The Life and Legacies of an Untimely Prophet*, which attends in more detail to locating the *Essay*, particularly to adding the temporal question about the precise social, political, and economic upheaval wracking England in the first half of 1798 as Malthus wrote his work to James's spatial insight into the importance of where Malthus wrote.[31] Taken together, these projects begin the process of grounding the universalizing rhetoric of laws in the time-space coordinates of their construction.

Similar work locating Malthus's rhetoric has also begun to attend to the text of the *Essay*. Mayhew has looked at the sources of geographical information on which Malthus drew to build his global surveys of the operation of the principle of population, showing the extent to which he used data selectively to reinforce predetermined conclusions.[32] Far more comprehensively from a postcolonial perspective, Bashford and Chaplin have undertaken to show Malthus's many interweavings with empire and the knowledge it produced. They chart in detail how Malthus selectively excerpted suitable information from a wide array of travel books to build his universal argument in the Great Quarto of 1803, ignoring material that was inconvenient. They also show that Malthus was closely connected to the British Empire, not just through working for the East India College in Haileybury from 1805 but also through personal financial dealings for his extended family. Bashford and Chaplin show the extent to which Malthus as an imperial citizen built a textual message that elided the darker side of empire and underplayed the rational strategies of population control and adjustment that travelers had reported among those purportedly at earlier stages of civilization.[33] Taken together, these analyses show the extent to which the universalizing rhetoric of Malthus's conjectural history relied on selective use of the available information. They suggest the limitations in Malthus's field of vision as he constructed his argument and show the "darker side" of Enlightenment savants, thereby grounding his purportedly universal knowledge claims in the culture of his age.

The purpose of the present chapter is to add to this critical tradition grounding Malthus's achievement. The aim is to achieve this by a spatial hermeneutic addressing the text of the *Essay* in its various iterations, not by a "hermeneutics of suspicion" that seeks to uncover the contextual elisions in Malthus's work. The argument will be that James's insight into Malthus's comment about "rosy cherubs" may be extended to show that the *Essay* betrays in its text and argumentation the importance of where Malthus wrote to what he wrote.

The Local and the Autobiographical in the First Edition of the *Essay* (1798)

For all the rhetoric of universal laws shot through the text of Malthus's *Essay* there is a set of traces of the autobiographical position in which Malthus found himself in 1798 and of the location in which he worked. Malthus's life story at this time is reasonably clear thanks to James's work. Having completed his studies in Cambridge, Malthus was living once more with his parents in Surrey and, by local family connections, had been given the curacy of Okewood.[34] He was also attempting to start a career as a writer about political affairs, having penned a tract called *The Crisis* in 1796 that would only reach print in fragmentary form in obituaries for him by William Otter and William Empson.[35] The "peak fear" that was 1798 for the British church and state responding to both revolutionary incendiarism and rural unrest in the face of rising food prices provided the context in which Malthus wrote the *Essay* that made his name.[36]

The *Essay* itself appeared with a preface dated June 7, 1798. It was published anonymously by Joseph Johnson of St Paul's Churchyard. Given that the *Essay* was immediately read as a significant counterrevolutionary tract in the context of "peak fear," it might strike one as odd that it was published by Johnson, a man who had been on government lists as a "suspect person" of potentially revolutionary sympathies as early as 1793 and who had published translations of an array of French radicals, most notable among whom was Condorcet, whose *Outlines* would, as we have already seen, be the subject of Malthus's ire about irrationalism.[37] Johnson was arrested by Pitt's government for purportedly seditious publishing only five weeks later, on July 17, 1798.[38]

Why did Johnson publish Malthus? The answer to this question almost certainly relates to a local connection between Johnson and Malthus in the form of their mutual acquaintance with rational dissenter Gilbert Wakefield, who had published with Johnson and who was Malthus's teacher. Wakefield had ceased to publish with Johnson but was in touch with him still. It was

Johnson's alleged publication of a tract by Wakefield (which he had, in fact, only ever sold for another publisher) that led to Johnson's trial later in the summer of 1798. It seems likely that Wakefield acted as the conduit between Johnson as established publisher and Malthus as neophyte author. It is likely that Johnson seized on this opportunity because he enjoyed publishing both sides of an argument and because the core of Johnson's interests was in dissenting literature, something to which the final two chapters of the *Essay* were perilously close for an Anglican clergyman, this doubtless reflecting the rational Anglican contexts in which Malthus had been educated. James Raven, in his account of the geographies of London publishing, notes the extent to which title pages, by flagging a publisher's location, "gave permanence to the association between declared author, printer, publisher and place," and that furthermore imprints could become associated with particular political and religious positions.[39] The disjunction between Malthus's publisher and the political position expressed in the *Essay* can be exaggerated: as publisher, Johnson was cast as a dangerous radical by his Pittite opponents, but even his actual position as a radical dissenter was at some remove from Malthus's Anglicanism. The disjunction between the ideology of Malthus's title page and his argument, unusual in this era as Raven makes clear, can be explained, then, by the centrality of the networks of kinship and tutelage by which Malthus was connected to Johnson via Wakefield. In short, that Malthus was published by Johnson seems to relate to the spaces of life in which he had been educated.

Turning from the context of the *Essay*'s publication to the work itself, we can identify three different strands in the text that disclose the local circumstances of its production: the locative and occasional, the autobiographical, and the local. By the locative, we mean elements in the text that bespeak the time and space of its production, that show occasional elements in its emergence and thereby belie the universalizing rhetoric deemed definitive of nascent social science. The *Essay* is shot through with such moments and opens with one in its preface, where Malthus notes that "the following essay owes its origin to a conversation with a friend" (1798, preface: i). The friend in question was Malthus's father, Daniel, which immediately locates an origin for the *Essay* in the family home in Albury, although no further details are offered.[40] The preface further notes that Malthus had wanted to make his argument more complete, "but a long and almost total interruption, from very particular business" (1798, preface: i), had led him to put the argument out as it was rather than delaying still further. This again draws the attention of the reader away from the argument and toward the occasional circumstances in which its author had worked. Samuel Johnson famously noted that "in lapidary inscriptions a man is not upon oath."[41] A similar argument might be made in terms of generic conventions for the role of the preface in Malthus's time: it was a textual space in which the writer was allowed to disclose the "back stage" circumstances of a work's composition, and, as such, occasional disclosures of this ilk were warranted. Yet Malthus, in fact, continues to disclose the occasional and locative in the text of the *Essay* as well. In opening his controversial final two chapters, for example, where Malthus laid out the theodicy he viewed as undergirding the principle of population, he added a note that he had wanted to offer a far more complete philosophical-cum-theological discourse but that "a long interruption, from particular business, has obliged me to lay aside this intention" (1798, ch. 18: 124). We do not know what this twice-mentioned interruption was. But it does pull the discourse of the *Essay*'s most far-reaching general claims about the functioning of the universe and the role of the principle of population in creating a state of human trial upon earth back into the realm of the author and the circumstances of his life at the point of composition. Malthus also alludes at two points to his location and his inability to access books with which he was familiar (presumably during his time in Cambridge). He notes in chapter 6 that he would quote from Ezra Style's pamphlet on population growth in North America but has had to take his information from Richard Price, "not having Dr Styles's pamphlet . . . by me" (1798, ch. 6: 40, note). He similarly acknowledged that his framing of an argument from Locke in chapter 18 is correct "if I recollect" (1798, ch. 18: 125), by implication again blaming this lack of scholarly precision on his inability to access books from his authorial location.[42]

If the *Essay* acknowledges its occasional and located space of production in both prefatory comments and marginal textual asides, it further discloses the moment in time in which it was produced. One element of this is well known: the *Essay* was a reaction to the utopian fervor reverberating in England from the French Revolutionary moment through the writings of Godwin and Condorcet, something Malthus notices right at the beginning of his formal argument in chapter 1 in reference to "that tremendous phenomenon in the political horizon the French revolution" (1798, ch. 1: 5).[43] The text reverts to this moment on a number of occasions and most notably in the chapters that engage with the ideas of Godwin and Condorcet directly, chapters 8 to 15. Far less noted have been those other moments where the text discloses the moment in time at which it was produced. The most important of these comes in chapter 7 where Malthus addresses the question of when an increase in population can be a sound socioeconomic outcome. He therein offers a more temporally specific aside that he can "entirely acquit Mr Pitt of any sinister intention in that clause of his Poor Bill which allows a shilling a week to every labourer for each child he has above three" but notes that such a policy would be disastrous. The bill in question was a reform of the Poor Laws that Pitt had proposed in 1797 but in which he was soundly defeated.[44] In referring to it, Malthus obviously locates his argument in terms of the moment of its production.

Malthus refers to several contemporary figures in English culture and society in asides that also locate the *Essay*. In condemning Godwin's modes of philosophical argument by aligning them to "the assertions of the prophet Mr Brothers" (1798, ch. 12: 83), for example, Malthus drew on the widespread panic-cum-derision with which Richard Brothers's alleged prophecies, including that of King George III's death, had been greeted. Brothers had been declared insane for this prophecy in 1795 and so was effectively neutralized as a source of fear.[45] Malthus aligned the very different force of Godwin's reasoning with Brothers's to demean it and to deny its force in a way that resonated with the historical moment. Finally, and in the same chapter, Malthus engaged with Godwin's claim that the power of the mind could overcome bodily discomforts, agreeing with this but only to a limited degree. In so doing, Malthus commented that different bodily habituations explained much of this, adding that "Powel, for a motive of ten guineas, would have walked further probably than Mr Godwin, for a motive of half a million" (1798, ch. 12: 80). Malthus here made a very located temporal reference that has less resonance today: it refers to Foster Powell, whose feats of long-distance competitive pedestrianism galvanized London until his death in 1793.[46] Intriguingly, Powell was buried in St. Faith's in St. Paul's Churchyard, a location directly opposite the print house of Joseph Johnson, Malthus's publisher.

Taken together, these comments from the text of the *Essay* reveal the occasional nature and moment of its production, grounding its universalizing language of axioms and propositions in a far more precise historical geography. This insight can be taken further by noting the extent to which the book also embodied the autobiographical location of Malthus himself. Reading the text of the *Essay* with an awareness of Malthus's life story at the moment of its composition, it can be noted that his personal position was reflected in several places, albeit clothed in a generalized language of universal observation. In his discussion of the idea of a preventive check to population growth being exerted by human reason, for example, Malthus drew a general picture of "a man of liberal education, but with an income only just sufficient to enable him to associate in the rank of gentlemen" (1798, ch. 4: 26). He suggested this situation would lead such a person to delay marriage as it would result in poverty for himself and his family such that they would have to "rank . . . with moderate farmers, and the lower class of tradesmen." It is hard not to see in this portrait the generalization of Malthus's own social position as he struggled to build a career and social position for himself in 1798 as a humble curate. Malthus only married in 1804 at the late age of 38 when he had a far more secure income. This folding of the autobiographical into a general language of social analysis is repeated toward the close of the *Essay* in chapter 18, where, as part of his theodicy, Malthus argues that "the middle regions of society seem to be best suited to intellectual improvement" (1798, ch. 18: 128), thereby locating social value in the class he himself inhabited. In the same vein, Malthus went on to add that leisure was inimical to improvement, evidence being that "talents are more common among younger brothers, than among elder brothers" (1798, ch. 18: 129), a position of comfort to him as a younger brother struggling to make ends meet in the middling sorts and with no prospect of any major inheritance. In addition to these autobiographically inflected generalisms about social mores, Malthus also offered one more portrait in the *Essay* that was directly autobiographical. In his previously noted discussion of Godwin's arguments about the relation of the mind to the body, Malthus drew on his own experiences when out shooting:

> When I have taken a long walk with my gun, and met with no success, I have frequently returned home feeling a considerable degree of uncomfortableness from fatigue. Another day, perhaps, going over nearly the same extent of ground with a good deal of sport, I have come home fresh, and alert. The difference in the sensation of fatigue upon coming in, on the different days, may have been very striking, but on the following mornings I have found no such difference. I have not perceived that I was less stiff in my limbs, nor less footsore, one the morning after the day of sport, than on the other morning. (1798, ch. 12: 81)

We know that Malthus was a keen hunter as he wrote of this frequently in his letters home from Cambridge.[47] It is notable that Malthus chose to pen this explicitly autobiographical portrait in the texture of a general argument about the universal laws of society. That he was willing to fold autobiographical and occasional information and asides into his general argument does not bespeak a "lapse" in the universality of his argument but is better seen as aligned with an intellectual position about the centrality of local or grounded knowledge to the construction of social knowledge.

Malthus in the texture of his *Essay* builds an argument about local knowledge as warranted and reliable knowledge that has gone all but unnoticed due to the emphasis on universal argumentation. To uncover this argument, we can revert to the "local" moment in the *Essay* identified by Patricia James in Malthus's comment on country laborers not being the "rosy cherubs" of fiction. This comment comes in the section of the *Essay* that rounds out Malthus's conjectural history of the principle of population as it has manifested itself across time and space in all stages of social development. Chapters 4 and 5 address modern societies to identify how the principle of population works. It is there that Malthus mentions the "rosy cherubs" as part of an argument that has considerable sweep in it, surveying from Caesar to China, but that ends up being far more precisely attentive to the local population dynamics Malthus himself could observe in the rural Surrey of 1798. Opening these chapters, Malthus noted that, looking at "the most civilized nations . . . we shall be assisted in our review by what we daily see around us, by actual experience, by facts that come within the scope of every man's observation" (1798, ch. 4: 23). True to this comment, for all that Malthus draws on travelers' accounts and statistical information about the demographic dynamics of European nations and North America, he also relies upon personal eyewitnessing, such that the local is imbricated into his picture of the demography of civilized nations. In the spirit of the conjectural history that frames his argument, Malthus suggests that "observations made [in England] will apply with but little variation to any other country where the population increases slowly"—that is, where preventive as well as positive checks exist (1798, ch. 4: 26).

It is immediately after this comment that he generalizes his personal situation in the aforementioned passage about "a man of liberal education" as he surveys the operation of the preventive check in modern societies, looking at how the same check operates on "the sons of tradesmen and farmers" and "the labourer who earns eighteen pence a day," in turn (1798, ch. 4: 27), before rounding out his survey with "the servants who live in gentlemen's families" (1798, ch. 4: 28). This "sketch of the state of society in England" (1798, ch. 4: 28) appears to relate to Malthus's own observations in a rural environment, notably as it does not include in its social hierarchy those in commerce, guilds, or industry. This impression is reinforced when Malthus turns to the operation of the positive check to population in advanced societies in chapter 5. There, Malthus notes that the operation of a positive check in the form of infant mortality has been obvious in urban environments due to "unwholesome habitations and hard labour," and has been apparent thanks to the statistical "bills of mortality" collected in towns and studied from John Graunt's *Observations on the Bills of Mortality* (1662) onward (1798, ch. 4: 29). While Malthus cannot offer statistical information about infant mortality in the countryside, he does suggest that while "it may not prevail in an equal degree in the country," nevertheless it is higher among the poorer rural laborers than among "the middling and higher classes" (1798, ch. 5: 29). It is on the depiction of rural poverty, not the better-known and more copiously evidenced questions of urban mortality, that Malthus focused in the *Essay* precisely because that is what he himself had been able to observe in his situation in 1798. It is here that the comment about "rosy cherubs" is offered with its intimation of autopsy on Malthus's part:

> The sons and daughters of peasants will not be found such rosy cherubs in real life, as they are described to be in romances. It cannot fail to be remarked by those who live much in the country, that the sons of labourers are very apt to be stunted in their growth, and are a long while arriving at maturity. Boys that you would guess to be fourteen or fifteen, are upon inquiry, frequently found to be eighteen or nineteen. And the lads who drive plough, which must certainly be a healthy exercise, are very rarely seen with any appearance of calves to their legs; a circumstance, which can only be attributed to a want either or proper, or of sufficient nourishment. (1798, ch. 5: 29–30)

For all the enlightened rhetoric of an "equal, wide survey," then, Malthus, in fact, focuses on the rural England that he had witnessed to build his picture of the social demography of civilized societies on the assumption that it can be transferred to other nations at the same stage of societal development.[48]

This reliance on the observed and the local was not just an empirical position. It was undergirded in Malthus's argument by a methodological-cum-political position. In conventional fashion for eighteenth-century writers, Malthus contrasted Newton with Descartes in seeing the former as "grand and consistent" because his arguments were "founded on careful and reiterated experiments" (1798, ch. 9: 59).[49] As with Newton in the natural world, to argue in the social sphere for realities that had never been reliably observed amounted to "an end of all human science. The whole train of reasonings from effects to causes will be destroyed. We may shut our eyes to the book of nature" (1798, ch. 9: 59). Yet this was exactly what, in the revolutionary moment, Malthus charged his utopian opponents with doing, seeing in Godwin and Condorcet's social science "the present rage for wide and unrestrained speculation" that amounted to "a kind of mental intoxication" (1798, ch. 9: 60, note). Malthus wanted to juxtapose his Newtonian social science to Cartesian "wild and eccentric hypotheses" (1798, ch. 9: 59), but in this binary his Newtonianism meant not just the quest for universal laws but also their grounding in specific, local observations of the social world in order to avoid the descent into unwarranted speculation. This is exactly why his analysis, most notably in chapters 4 and 5 but elsewhere also, had been so clearly tied to his own social and geographical situation and observations. It is noteworthy in this context that Malthus's most controversial policy suggestion—elimination of the Poor Laws as they "create the poor which they maintain" (1798, ch. 5: 33)—emerged directly after the observationally driven reportage about the rural poor as stunted rather than cherublike as a way "to remedy the frequent distresses of the common people" (1798, ch. 5: 30) that he had just chronicled. For Malthus, then, a grounded local observational knowledge was the way to construct Newtonian laws of the social world and to scotch the seductive fantasies of utopian reason. For this reason, the *Essay* by its very nature exhibited local detail in its construction of a universal rhetoric.

Observation and Local Knowledge in the "Great Quarto," 1803–26

Five years after Malthus appeared on the public stage, a massively expanded second edition of his *Essay*, often known as the "Great Quarto," appeared. Being four times longer than its 1798 predecessor, Malthus himself noted "in its present shape it may be considered as a new work" (GQ 1:2).[50] The Great Quarto would go through further iterations in 1806, 1807, 1817, and 1826 but always as modifications of the argument produced in 1803. From 1803, Malthus's authorship was noted on the title page, Malthus being styled "Fellow of Jesus College, Cambridge" in Johnson's new edition. From 1817, after Johnson's death in 1809 and the subsequent collapse of his imprint in 1814, Malthus became an author with John Murray, the leading publisher of the age.[51] The last two editions bore witness to Malthus's aggrandizement as an author and member of select social circles, offering that he was "Late Fellow of Jesus College, Cambridge, and Professor of History and Political Economy in the East India College, Hertfordshire" (as the 1826 edition has it). The massive expansion of the *Essay* and the success of its author might have effaced the local and ephemeral aspects of the work that we have uncovered in the first edition, something which the preface to the Great Quarto made likely in referring to its 1798 predecessor as "written on the spur of the occasion, and from the few materials that were within my reach in a country situation" (GQ 1:1).

The truth is rather more complex. It is more accurate to say that the balance between autobiographical and local material changes in the Great Quarto editions from that we have uncovered for the first edition, but that the local, grounded nature of Malthus's analysis of social, economic, and demographic material remains and is, if anything, expanded. This is foreshadowed even as Malthus distances the project of 1803 from the occasional context of 1798 in his very preface. Here, Malthus notes that he could have merely attended to the logic of his argument to make it "an impregnable fortress" but that his real interest is in the policy outcomes the principle of population necessitates, and that in analyzing these by direct observation he will "have opened the door to many objections" (GQ 1:2–3). In his first edition for John Murray in 1817, Malthus reiterated this position in a new preface, saying that adjustments were made to ensure the continued "application of the general principles of the Essay to the present state of things" (GQ 1:7). Where in 1803 the *Essay* had been "published at a period of extensive warfare, combined, from peculiar circumstances, with a most prosperous foreign commerce," economic depression in the wake of the Napoleonic Wars meant "a period of a different description has succeeded, which has in the most marked manner illustrated its [the *Essay*'s] principles, and confirmed its conclusions" (GQ 1:6). While Malthus laid bare the occasional origins of the 1798 *Essay*, in the same prefatory material he continued to assert the need for any general analysis of society to be grounded in observations derived from, and policies aimed at, the specific time and space in which it was constructed. Localism would remain in the Great Quarto.

The generalized autobiographical disclosures we have noted in the 1798 *Essay* were by and large dropped from 1803 onward. We no longer see Malthus out hunting, and the complete excision of the controversial theological elements of the 1798 edition saw the removal of Malthus's speculations on the middling sort and on younger siblings within that class. Malthus retained his general comments on the man of "liberal education" and his unwillingness to marry until he can do so without descending the social ladder in all the editions of the *Essay*. This was now placed in book II of the Great Quarto's examination "Of the Checks to Population in England" (GQ 1:250): the title reflecting the local origins of this observation. In refusing to generalize it to other civilized nations, Malthus thereby reversed the position on the generalizability of local knowledge he had espoused in chapter 4 of the 1798 *Essay*. The one obviously autobiographical element added to the Great Quarto came in a discussion of marriage in book IV, chapter 9. In 1803, while Malthus remained a bachelor, this discussion criticized the social approbation attached to the mere fact of marriage, deeming it irrational "that a giddy girl of sixteen should, because she is married, be considered by the forms of society as the protector of women of thirty" (GQ 2:150), as this discernibly weakened the preventive check to population increase. In 1806, this entire section was removed as Malthus himself had married in 1804.[52] While Malthus might have removed this passage due to those frequent lampoons of him as an advocate of spinsterhood that would peak in the satirical novel *Melincourt* (1817) by Thomas Love Peacock, it is noticeable that this adjustment coincides so closely with his change in marital state. The autobiographical, then, was all but excised from the Great Quarto editions, this type of locative identifier presumably being deemed inappropriate to the context and ambitions of Malthus's expanded argument.

As the autobiographical element of Malthus's project receded, however, the local and observational side expanded. The main reason for the massive increase in the bulk of the *Essay* was Malthus's commitment to tracing the empirical evidence that supported his argument for a universally discernible principle of population. As he put it, "a more practical and permanent interest" would be added to his work by a greatly expanded "historical examination of the effects of the principle of population on the past and present state of society" (GQ 1:1). As such, two massive books culled from historical accounts and travelers' narratives replaced the four-chapter, conjectural-historical tour of the globe offered in 1798. In this context, unsurprisingly given his commitment to personal observation as the basis of a reliable human science, Malthus from 1803 drew on his own knowledge and experience to a greater extent than five years previously. This becomes most obvious in book II of the Great Quarto, which, addressing "checks to population in the different states of modern Europe," amounts to an empirically grounded reworking of the observational argument we have traced in chapters 4 and 5 of its 1798 predecessor.

Opening book II, Malthus noted importantly for his approach that "political calculators have been led into the error of supposing that there is, generally speaking, an invariable order of mortality in all countries; but it appears, on the contrary, that this order is extremely variable; that it is very different in different places of the same country and, within certain limits, depends upon circumstances which it is in the power of man to alter" (GQ 1:148). As with the later section concerning population checks in England, in a sense this reversed the position adopted in 1798, wherein he argued that the evidence he could observe in England was a reasonable proxy for all civilized nations. Now not only did Malthus argue that different countries would exhibit different demographic dynamics but that different parts of different countries would as well, this demanding a more geographically nuanced approach. Book II of the Great Quarto provided this in its country-by-country survey of modern Europe. This did not mean that Malthus relinquished the emphasis on direct eye witnessing from 1803 onward. On the contrary, the five years since the first edition of the *Essay* had seen Malthus travel to Norway, Sweden, and Russia in 1799 and to Switzerland during the cessation of hostilities in the Napoleonic Wars marked by the Peace of Amiens in 1802.

The desire to canvass more nations but to do so on the basis of direct evidence is reflected in the rather eccentric "tour" of European demography that is enacted in book II of the Great Quarto. Most British geographical descriptions of the age either began with their home nation and then moved outward to continental Europe or followed a geographically predictable course, normally from West to East.[53] Book II of the Great Quarto describes an arc, starting with the very infrequently visited and poorly known northern nations of Norway, Sweden, and Russia, looping down to Switzerland and France, before finishing by moving north again to England, Scotland, and Ireland.[54] It is possible to argue that Malthus began with Norway as the most advanced demographic regime known to him, framing its history as he did as showing that "the positive checks to its population have been so small, the preventive checks must have been proportionably great" (GQ 1:149). It could also be deemed a logical starting place: its small population laid bare the economic and social problems created by a large, dependent class that was the main policy outcome Malthus sought to enforce. He concluded that Norway was "almost the only country in Europe where a traveller will hear any apprehensions expressed of a redundant population," something he ascribed to "the smallness of the population altogether, and the consequent narrowness of the subject" (GQ 1:157). Yet this would not explain why Malthus's European survey neglected Eastern, Mediterranean, and Iberian Europe, only addressing them as ancient and less civilized areas in book I.[55] It seems far more plausible, then, to ascribe the scope and shape of Malthus's circuit of modern nations in book II to the determining factor of places with which he was personally acquainted.

Analysis of how Malthus wrote about modern European nations confirms this impression as he clearly and persistently interwove material taken from his own observations during his travels into the texture of his argument. In editing the Great Quarto, Patricia James rightly notes that in many places "Malthus . . . quotes himself almost verbatim" from the travel diaries he had kept (GQ 1:149n3).[56] Some of this material was frankly irrelevant to his argument, such as his note about the "strikingly picturesque" valleys in Sweden and his particular appreciation of "the principal road from Christiania to Drontheim" (GQ 1:159n9). Sometimes information Malthus had gleaned on his travels, while relevant to his argument, seemed to bulk disproportionately large because he had clearly found it interesting as he traveled. This is most notable in the chapter on Russia where the discussion of a single foundling hospital in St. Petersburg takes up four of the ten pages Malthus addresses to the country as a whole (GQ 1:173–76).[57] In general, Malthus skillfully emphasized his personal observations within a broader, more generalized analysis of the demographic regimes of the nations he had visited. In his analysis of Sweden, Malthus draws on statistical information by Wargentin to analyze the mortality regime and on data from the *Memoires du Royaume de Suede* to analyze food imports but uses his own observations from 1799 regarding agriculture, climate, soil, and food yields and their relation to demographic dynamics, reverting in his discourse to the first person and grounding his ideas as based "on observing many spots of this kind both in Norway and Sweden" (GQ 1:163). He was candid about the extent of his personal eyewitnessing in different places. Where Malthus spent some considerable time analyzing the relations of population, food, and policy in Sweden, Norway, and Russia, and interwove this into his arguments, he acknowledged in discussing Switzerland the more occasional nature of his knowledge as he had been there on an "excursion" (GQ 1:226) and that the "disturbed state of the country" (GQ 1:224) due to the chaos of conflict fomented by Napoleon meant he could not collect much reliable information. It was in his discussion of Switzerland that he included a very anecdotal account, also recorded by his future wife, Harriet Eckersall, of a peasant guide in the Jura region who ascribed "the misery and starving condition of the greatest part of the society" to those who "still continued to marry and to produce a numerous offspring which they could not support" (GQ 1:227).[58] Clearly, Malthus's traveling party had been amused by this echo of his theory in the 1798 *Essay*, and Malthus kept this anecdotal aside in place throughout the five editions of the Great Quarto.

The commitment to direct observation as the reliable foundation for the study of society that Malthus had forged in the context of revolutionary fervor stayed with him in the Great Quarto editions of the *Essay*, the local elements from 1798 being transformed into a more directly observational framework of demographic autopsy. It was this that explains the otherwise eccentric organization of book II and the distinctive implicit definition of the geography of the nations of "modern" Europe. To an extent that has gone unnoticed, the grounding of purportedly general observations about demography continued to be in the resolutely local observations of the author. The other place where Malthus retained a sense of the local or autobiographical in the Great Quarto edition of the *Essay* was in his defense of his controversial proposal to abolish the Poor Laws. In a chapter added in 1817 (GQ 3, ch. 7), Malthus compiled a long series of responses to his critics, in all of which he writes repeatedly in the first person about "what I have really proposed" (GQ 1:374) as opposed to the caricatures of his critics.

The Great Quarto shifts the defense of his proposals away from being grounded in his knowledge of the English countryside and the miseries of its less-than-rosy cherubs, as in 1798, toward the comparative insights he garnered as a traveler. Malthus notes that such a system of parish relief would "convulse the social system" in Sweden and that in France "the landed property would soon sink under the burden" (GQ 2:146). Only Holland, for Malthus, had successfully run a system of parish relief, and this is because of its specific combination of "extensive foreign trade . . . numerous colonial emigrations," and the extreme unhealthiness of the country, which kept mortality high (GQ 2:146). The most important proof of the case came with Norway, where "notwithstanding the disadvantage of a severe and uncertain climate, from the little that I saw in a few weeks residence in the country . . . I am inclined to think that the poor were, on the average, better off than in England" (GQ 2, 146) due to the self-reliance inculcated by the lack of a system of parish relief and the resultant power of the preventive check to population increase. A parallel position was adopted in rebutting those who argued—as would Hazlitt, Cobbett, Southey, and others—that delayed marriage as advocated by Malthus would only lead to increased vice, notably in the form of prostitution: "the instances of Norway, Switzerland, England and Scotland, adduced in . . . this Essay, show that, in comparing different countries together, a small proportion of marriages and births does not necessarily imply the greater prevalence even of this particular vice" (GQ 2:222). Malthus's rhetorical location, then, shifted from that of a middle-class observer living in the English countryside in the *Essay* of 1798 to, in the Great Quarto editions, that of a traveler acting as a comparative analyst of population and resource dynamics across (his autobiographically shaped definition of) modern Europe. This shift buttressed his policy prescriptions about delayed marriage, poor relief, and the causes of poverty.

Conclusion: Malthus, Space, and Social Science

If, as Malthus believed, the scientific analysis of societies had to be built out of direct observation to avoid descending into the realm of utopian fantasy, it would by that very fact have to be geographically grounded as well in the local and peregrinatory experiences of an observer or their sources. This belief was enacted in different ways in the first edition of the *Essay* in 1798 and its successor project, the Great Quarto of 1803–26. In 1798, the *Essay* was tied to Malthus's autobiographical position, but more importantly to his observation of England's rural poor, whose lot was clearly a motivating force both for the writing of the book and for its prescriptions with regard to the Poor Laws. The revolutionary moment of 1798 was obviously important to the *Essay*, as was the desire to contribute to a general conjectural-historical account of the stages through which societies developed. This should not blind us to the geographical, spatial, and local grounding of the *Essay*.

The rhetorical and observational location of Malthus's work changed considerably for the 1803 edition and its successors in good part because Malthus had now traveled more widely and could thereby use a far less parochial range of observations to support his arguments. It is noticeable that the geographical survey of Europe in the Great Quarto is so closely tied to Malthus's own acts of eye witnessing. For all the evidentiary buttressing of his arguments by data from censuses, scholars, and academies, the structure of Malthus's textual tour of modern Europe follows his own peregrinations, while the argument offered in that tour insistently returns to observations and anecdotes Malthus recorded as he traveled intensively in the years between the two versions of his *Essay*. In both iterations, then, the *Essay on the Principle of Population* was a very grounded work of local knowledge, something that has been obscured both by its universalizing rhetoric of laws, axioms, and principles and by a historiography looking to define the birth of the social sciences precisely through the construction of such a rhetoric.

The present argument has sought to recover the local, grounded, and geographically identifiable elements in Malthus's argument. The suggestion is that what is being called here a "spatial hermeneutic"—a reading strategy attentive to the geographical disclosures made by an author—can cut against the grain of received readings of canonical texts such as Malthus's *Essay* and of historiographical truisms that locate in the emergence of universalizing discourses about society the birth of modern social science. By tying together the "spaces of life" in which an author functioned with the spatial disclosures offered by their texts, new historical geographies of scholarship—scientific and otherwise—can be disclosed. The aim of such a spatial hermeneutic is not to belittle the achievements of Malthus or others as local or ephemeral but to show the inevitably local and grounded moorings of their inquiries.[59] This is neither a suspicious nor a deconstructive hermeneutic but one that sees grandeur in tracing the geographically identifiable contexts in which ideas were generated and through which they were built. One does not have to reverse the polarization of the local and the universal that has been applied in most historiography of the physical and social sciences as did William Blake in his celebrated rebuttal of Reynolds's *Discourses on Art*—"To Generalise is to be an Idiot; To Particularise is the Alone Distinction of Merit"—to hold the two in productive tension.[60]

Notes

1. David N. Livingstone, *Putting Science in Its Place: Geographies of Scientific Knowledge* (Chicago: University of Chicago Press, 2003), 1.

2. See in particular the series of volumes edited by Mikuláš Teich and Roy Porter for Cambridge University Press addressing a range of historical topics in national context, the first of which was published in 1981, the volume concerning the Scientific Revolution being published in 1992. For early studies of regional science, see Ian Inkster and Jack Morrell, eds., *Metropolis and Province: Science in British Culture, 1780–1850* (London: Hutchinson, 1983), and for colonial science, see James E. Mclellan III, *Colonialism and Science: Saint Domingue and the Old Regime* (Baltimore: Johns Hopkins University Press, 1992).

3. Adi Ophir and Steven Shapin, "The Place of Knowledge: A Methodological Survey," *Science in Context* 4 (1991): 3–22; and David Turnbull, "Reframing Science and Other Local Knowledge Traditions," *Futures* 29 (1997): 551–62.

4. David N. Livingstone and Charles W. J. Withers, eds., *Geography and Enlightenment* (Chicago: University of Chicago Press, 1999); David N. Livingstone and Charles W. J. Withers, eds., *Geography and Revolution* (Chicago: University of Chicago Press, 2005); and David N. Livingstone and Charles W. J. Withers, eds., *Geographies of Nineteenth-Century Science* (Chicago: University of Chicago Press, 2011).

5. Some of the scope of this work is helpfully canvassed in important review articles: see Richard Powell, "Geographies of Science: Histories, Localities, Practices, Futures," *Progress in Human Geography* 31 (2007): 309–29; Diarmid Finnegan, "The Spatial Turn: Geographical Approaches in the History of Science," *Journal of the History of Biology* 41 (2008): 369–88; and Diarmid Finnegan, *Geography of Knowledge*, Oxford Bibliographies (Oxford: Oxford University Press, 2013).

6. Compare J. G. A. Pocock, *Barbarism and Religion: Volume 1: The Enlightenments of Edward Gibbon, 1737–1764* (Cambridge: Cambridge University Press, 1999) with John Robertson, *The Case for the Enlightenment: Scotland and Naples, 1680–1750* (Cambridge: Cambridge University Press, 2005). The geographical elements of recent intellectual history are discussed in Robert J. Mayhew, "Geography as the Eye of Enlightenment Historiography," *Modern Intellectual History* 7 (2010): 611–27; Charles W. J. Withers, *Placing the Enlightenment: Thinking Geographically about the Age of Reason* (Chicago: University of Chicago Press, 2007); Charles W. J. Withers, "Place and the 'Spatial Turn' in Geography and History," *Journal of the History of Ideas* 70 (2009): 637–58; and Charles W. J. Withers and Robert J. Mayhew, "Geography: Space, Place and Intellectual History in the Eighteenth Century," *Journal of Eighteenth Century Studies* 34 (2011): 445–52. For a state-of-the-art collection, see also Paul Stock, ed., *The Uses of Space in Early Modern History* (New York: Palgrave, 2015).

7. Peter Galison, "Ten Problems in History and Philosophy of Science," *Isis* 99 (2008): 111–24, quote on 119.

8. Some work has addressed the role of the local in scientific activity: see David Chambers and Richard Gillespie, "Locality in the History of Science: Colonial Science, Technoscience, and Indigenous Knowledge," *Osiris* 15 (2000): 221–40; Lewis Pyenson, "An End to National Science: The Meaning and the Extension of Local Knowledge," *History of Science* 40 (2002): 251–90; and the section by Kapil Raj and Mary Terrall, eds., "Circulation and Locality in Early Modern Science," *British Journal for the History of Science* 43 (2010): 513–606. For a thorough survey of the sites of science, see Livingstone, *Putting Science in Its Place*, 17–86.

9. Livingstone, *Putting Science in Its Place*, 183; emphasis original.

10. Martin J. S. Rudwick, *The Great Devonian Controversy: The Shaping of Scientific Knowledge amongst Gentlemanly Specialists* (Chicago: University of Chicago Press, 1985); and Adrian Desmond, *The Politics of Evolution: Morphology, Medicine and Reform in Radical London* (Chicago: University of Chicago Press, 1989).

11. Patricia James, *Population Malthus: His Life and Times* (London: Routledge, 1979). The most significant biographical additions come from the material collected in John Pullen and Trevor Hughes Parry, eds., *T. R. Malthus: The Unpublished Papers in the Collection of Kanto Gakuen University*, 2 vols. (Cambridge: Cambridge University Press, 1997–2004).

12. David N. Livingstone, "Science, Text and Space: Thoughts on the Geography of Reading" *Transactions of the Institute of British Geographers* 30 (2005): 391–401, quote on 392; emphasis original.

13. See, in particular, David N. Livingstone, *Dealing with Darwin: Place, Politics, and Rhetoric in Religious Engagements with Evolution* (Baltimore: Johns Hopkins University Press, 2014). On reader reception and spatiality, see James A. Secord, *Victorian Sensation: The Extraordinary Publication, Reception, and Secret Authorship of* Vestiges of the Natural History of Creation (Chicago: University of Chicago Press, 2000), his *Visions of Science: Books and Readers at the Dawn of the Victorian Age* (Oxford: Oxford University Press, 2014), and Innes Keighren, *Bringing Geography to Book: Ellen Semple and the Reception of Geographical Knowledge* (London: I. B. Tauris, 2010).

14. See Livingstone and Withers, *Geographies of Nineteenth-Century Science*, chap-

ters by Livingstone, Withers, Alberti, and Kohlstedt. Similar strictures apply to the papers collected in Simon Naylor, ed., "Historical Geographies of Science," a special issue of the *British Journal for the History of Science* 38 (2005): 1–100, and Peter Meusberger, David Livingstone, and Heike Jöns, eds., *Geographies of Science* (Berlin: Springer, 2010).

15. This chronology drives the construction of the monumental efforts embodied in Theodore M. Porter and Dorothy Ross, eds., *The Cambridge History of Science: Volume 7: The Modern Social Sciences* (Cambridge: Cambridge University Press, 2003). See also Anthony Giddens, *Capitalism and Modern Social Theory: An Analysis of the Writings of Marx, Durkheim and Max Weber* (Cambridge: Cambridge University Press, 1971), and Johan Heilbron, *The Rise of Social Theory* (Cambridge: Polity Press, 1995).

16. See, for example, Geoffrey Hawthorn, *Enlightenment and Despair: A History of Sociology* (Cambridge: Cambridge University Press, 1976).

17. See Hawthorn, *Enlightenment and Despair*; Heilbron, *Rise of Social Theory*; and Christopher J. Berry, *Social Theory of the Scottish Enlightenment* (Edinburgh: Edinburgh University Press, 1997).

18. Ronald L. Meek, *Social Science and the Ignoble Savage* (Cambridge: Cambridge University Press, 1976), quote on 242. On conjectural history and the birth of social thought, see also Larry Woolf and Marco Cipolloni, eds., *The Anthropology of Enlightenment* (Stanford: Stanford University Press, 2007); Frederick G. Whelan, *Enlightenment Political Thought and Non-Western Societies: Sultans and Savages* (London: Routledge, 2009); and Frank Palmeri, *State of Nature, Stages of Society: Enlightenment Conjectural History and Modern Social Discourse* (New York: Columbia University Press, 2016).

19. Björn Wittrock, Johan Heilbron, and Lars Magnusson, "The Rise of the Social Sciences and the Formation of Modernity," in *The Rise of the Social Sciences and the Formation of Modernity: Conceptual Change in Context, 1750–1850*, ed. Johan Heilbron, Lars Magnusson, and Björn Wittrock (Dordrecht: Kluwer Academic Publishers, 1998), 1–33, quote on 22.

20. Robert Wokler, "The Enlightenment and the French Revolutionary Pangs of Modernity," in Heilbron, Magnusson, and Wittrock, *Rise of the Social Sciences*, 35–76, quote on 41; cf. Keith Baker, *Condorcet: From Natural Philosophy to Social Mathematics* (Chicago: University of Chicago Press, 1975), chapters 4 and 5.

21. See David Carrithers, "The Enlightenment Science of Society," in *Inventing Human Science: Eighteenth-Century Domains*, ed. Christopher Fox, Roy Porter, and Robert Wokler (Berkeley: University of California Press, 1995), 232–70, quote on 234.

22. Wokler, "Enlightenment and French Revolutionary Pangs," in Heilbron, Magnusson, and Wittrock, *Rise of the Social Sciences*, 44–45, quote on 45. See a similar periodization for the English case in Eileen Janes Yeo, *The Contest for Social Science: Relations and Representations of Gender and Class* (London: River Orams Press, 1997).

23. Carrithers, "Enlightenment Science of Society," in Fox, Porter, and Wokler, *Inventing Human Science*, 246 and 245.

24. Berry, *Social Theory*, 52; cf. Johan Heilbron, "Social Thought and Natural Science," in Porter and Ross, *Cambridge History of Science, Volume 7*, 40–56.

25. Malthus is discussed by scholars of conjectural history for his usage of its structure: see Meek, *Social Science*, 223–24, and Palmeri, *State of Nature, Stages of Society*, 61–69.

26. Koselleck does not discuss Malthus, but Foucault did in discussing the emergence of the concept of statecraft about population: Michel Foucault, *Security, Territory, Population: Lectures at the Collège de France, 1977–1978*, ed. Michel Senellart (London: Palgrave, 2007), 55–86.

27. Michael Donnelly, "From Political Arithmetic to Social Statistics: How Some Nineteenth-Century Roots of the Social Sciences Were Implanted," in Heilbron, Magnusson, and Wittrock, *Rise of the Social Sciences*, 225–39, quote on 233.

28. Thomas Robert Malthus, *An Essay on the Principle of Population* (1798), volume 1 in E. A. Wrigley and David Souden, eds., *The Works of Thomas Robert Malthus*, 8 vols. (London: Pickering and Chatto, 1986), 59. All future references to the 1798 *Essay* in the text are to this edition, by chapter and page number.

29. Dipesh Chakrabarty, *Provincializing Europe: Postcolonial Thought and Historical Difference* (Princeton, NJ: Princeton University Press, 2000).

30. James, *Population Malthus*, 43.

31. Robert J. Mayhew, *Malthus: The Life and Legacies of an Untimely Prophet* (Cambridge, MA: Harvard University Press, 2014), 60–65.

32. Robert J. Mayhew, "Malthus's Globalisms: Enlightenment Geographical Imaginaries in the Essay on the Principle of Population," in *Spaces of Global Knowledge: Exhibition, Encounter and Exchange in an Age of Empire*, ed. Diarmid Finnegan and Jonathan Jeffrey Wright (Aldershot: Ashgate, 2015), 167–83.

33. Alison Bashford and Joyce E. Chaplin, *The New Worlds of Thomas Robert Malthus: Rereading the Principle of Population* (Princeton, NJ: Princeton University Press, 2016). For more on Malthus and Haileybury, see Timothy L. Alborn, "Boys to Men: Moral Restraint at Haileybury College," in *Malthus, Medicine and Morality: "Malthusianism" after 1798*, ed. Brian Dolan (Amsterdam: Rodopi, 2000), 33–55.

34. James, *Population Malthus*, 34–45.

35. Ibid., 50–54. See [William Otter], *Memoir of Robert Malthus*, in T. R. Malthus, *Principles of Political Economy*, 2nd ed. (London: John Murray, 1836), xiii–liv, with *The Crisis* discussed on xxxv–xxxvi; and [William Empson], "Life, Writings and Character of Mr Malthus," *Edinburgh Review* 64 (1837): 469–506, with *The Crisis* discussed on 479–82.

36. See James, *Population Malthus*, 46–61; and Mayhew, *Malthus*, 60–65.

37. Helen Braithwaite, *Romanticism, Publishing and Dissent: Joseph Johnson and the Cause of Liberty* (London: Palgrave, 2003), 143, for suspect person and 93–94 for Condorcet. See also Leslie F. Chard, "Joseph Johnson: Father of the Book Trade," *Bulletin of the New York Public Library* 79 (1975): 51–82; and Leslie Chard, "From Bookseller to Publisher: Joseph Johnson and the English Book Trade, 1760–1810," *The Library* 32 (1977): 138–54.

38. On this, see Jane Smyser, "The Trial and Imprisonment of Joseph Johnson, Bookseller," *Bulletin of the New York Public Library* 77 (1974): 418–35. See also Gilbert Wakefield, *Memoirs of the Life of Gilbert Wakefield*, 2 vols. (London: J. Johnson, 1804), 2:115–61.

39. James Raven, *Bookscape: Geographies of Printing and Publishing in London before 1800* (London: British Library, 2014), 147.

40. For the friend as Daniel Malthus, see Otter, *Memoir*, xxxviii–xxxix.

41. James Boswell, *Life of Johnson*, ed., G. B. Hill and L. F. Powell, 6 vols. (Oxford: Clarendon Press, 1934–50), 2:407.

42. It is very hard to find a precise correlate of Malthus's recollection of Locke in Locke's writings: see Thomas Robert Malthus, *An Essay on the Principle of Population and Other Writings*, ed. Robert J. Mayhew (London: Penguin, 2015), 295n110.

43. The subtitle of the book also made clear this point of origination for the *Essay*: "With Remarks on the Speculations of Mr Godwin, M Condorcet and Other Writers."

44. See Mayhew, *Malthus Essay*, 289n40.

45. For the serious concerns about Brothers, see John Barrell, *Imagining the King's Death: Figurative Death, Fantasies of Regicide, 1793–1796* (Oxford: Oxford University Press, 2000), 504–50.

46. See Henry Wilson and James Caulfield, *The Book of Wonderful Characters* ([1869] Cambridge: Cambridge University Press, 2012), 106–9. See also more generally Robin Jarvis, *Romantic Writing and Pedestrian Travel* (London: Palgrave, 1997), chapter 1.

47. Pullen and Parry, *Unpublished Papers*, 1:27–60.

48. John Barrell, *English Literature in History, 1730–1780: An Equal, Wide Survey* (London: Harper Collins, 1983).

49. On debates about Descartes and Newton, see J. B. Shank, *The Newton Wars and the Beginning of the French Enlightenment* (Chicago: University of Chicago Press, 2008).

50. T. R. Malthus, *An Essay on the Principle of Population*, ed. Patricia James, 2 vols. (Cambridge: Cambridge University Press for the Royal Economic Society, 1989), 1:2. Further references to the editions of the Great Quarto *Essay* in the text are to GQ, by volume and page number.

51. On Johnson, see Braithwaite, *Romanticism, Publishing and Dissent*, 179–81. For Murray, see William Zachs, *The First John Murray and the Late Eighteenth-Century London Book Trade: With a Checklist of His Publications* (Oxford: Oxford University Press / The British Academy, 1998); and Humphrey Carpenter, *The Seven Lives of John Murray: The Story of a Publishing Dynasty* (London: John Murray, 2009).

52. James, *Population Malthus*, 162–64.

53. On these structuring devices in geographical writing, see Robert J. Mayhew, "Materialist Hermeneutics, Textuality and the History of Geography: Print Spaces in British Geography, c. 1500–1900," *Journal of Historical Geography* 33 (2007): 466–88.

54. Scandinavia was still very infrequently visited by English travelers at this time: see Brian Dolan, *Exploring European Frontiers: British Travellers in the Age of Enlightenment* (Basingstoke: Macmillan, 2000).

55. Mayhew, "Malthus's Globalisms."

56. Patricia James, ed., *The Travel Diaries of T. R. Malthus* (Cambridge: Cambridge University Press for the Royal Economic Society, 1966).

57. Malthus's *Travel Diaries* stop short of his visit to Russia in 1799, but this material in the Great Quarto makes it almost certain that he kept a diary throughout the journey of which only the earlier sections have survived.

58. Harriet Eckersall's journal entry relating to this incident can be found in James, ed., *Travel Diaries*, 296–98.

59. In this regard, see also Simon Schaffer, "Newton on the Beach: The Information Order of the *Principia Mathematica*," *History of Science* 47 (2009): 436–52.

60. David V. Erdman, *The Complete Poetry and Prose of William Blake*, 2nd ed. (New York: Random House, 1982), 641.

CHAPTER TWO

Revisiting Belfast

Tyndall, Science, and the Plurality of Place

DIARMID A. FINNEGAN

Places, David Livingstone has suggested, "are constituted by the activities of the actants occupying them." At the same time, those same agents are "constrained by the system of human interactions pertaining in specific locations."[1] Doreen Massey has offered a comparable working definition. Places, she proposed, are composed of changing "mixtures of influences" that create and re-create unique "constellations of social relations."[2] As Massey further argued, places, at once distinctive and dynamic, are formed not only through dense local relations but also by multiple connections with a wider world. These thoughts on place as a concept have been developed and redrawn by others.[3] Such conceptualization points to the opportunities and challenges of mobilizing place as an analytical category and set of grounded realities.[4]

In this chapter, the place brought into view is nineteenth-century Belfast, a town (and, from 1888, city) that, like any other urban center, cannot be reduced to a single frame or understood as a stable agent that, in any straightforward way, shaped the production and promotion of scientific knowledge. Here, it is viewed through the prism of three events and one central character, the Irish-born physicist John Tyndall. I reconstruct three visits Tyndall made to the town in 1852, 1874, and 1890, and use each as an entry point into different accounts of Belfast as a place where science was mobilized for a variety of often conflicting ends.

This empirical focus raises further conceptual concerns related to Belfast as a place where scientific knowledge was communicated and contested. What is brought to view is clearly not an exhaustive account of Belfast as a total urban environment. Rather, it is the Belfast that shapes and is shaped by Tyndall's various interventions and interactions during his several visits that is of concern. In particular, spoken communication—both formal and informal—was among the critical practices mobilized by Tyndall and his interlocutors. It is useful, then, to view the Belfast that came to light through Tyndall's visits as, in significant ways, related to particular spaces of conversation and communication. There are, in other words, good reasons to think of Belfast as a "place" made in terms of relations between speech and space or, as Livingstone puts it, "location and locution."[5]

Tyndall's second visit in 1874 as president of the British Association for the Advancement of Science (BAAS) has attracted considerable scholarly attention. Of all his visits, this one has been understood as an event that significantly altered attitudes to evolutionary science and incited reaction across a wider sphere of intellectual and cultural life. Recent accounts of Tyndall's Belfast address have taken both aspects seriously. David Livingstone has subjected the responses of Belfast's Presbyterians to Tyndall's address to detailed scrutiny.[6] In Livingstone's account, Tyndall's address, in combination with reactions to it, helped produce an enduring antagonism toward Darwinism among Belfast's Calvinists. For leading Presbyterians, Belfast became a place where it was difficult to offer positive or accommodating appraisals of evolutionary ideas. In others' work, the reports and replies found in periodicals and other published media have been intensively investigated.[7] In this more widely dispersed textual form, favorable as well as critical responses to Tyndall's address were aired. To paint in broad strokes, among Belfast's Presbyterian community the reaction was predominantly if not wholly negative. In the more scattered periodical literature, the responses included more supportive commentary. This may leave the impression that Belfast, as a place strongly shaped by conservative religious forces, was uniformly opposed to Tyndall and for the same theological reasons.

This impression does not do justice to Livingstone's more carefully stated position that Tyndall faced opposition from *Presbyterian* Belfast. This qualification accords with the understanding of "place" as something that is already "there" but also something that is made and remade through the actions and reactions of particular agents. This chapter continues this line of thought by illustrating and examining Tyndall's encounters with Belfast on three separate occasions. I offer a diachronic as well as a synchronic analysis of Belfast as a place where certain kinds of scientific claims were articulated and appropriated in ways at once informed by, and transformative of, aspects of the town's changing religious and political culture. Reconstructing the three visits provides material to qualify a view of Tyndall's reception in Belfast as uniformly oppositional. It also provides the opportunity to uncover the ways in which Tyndall's various platform pronouncements became rapidly entrenched within the political and social interests of different constituencies.

Science and Religion in Early Victorian Belfast

On the afternoon of September 1, 1852, John Tyndall checked into his Belfast hotel on Joy Street. That evening, he walked to May Street Presbyterian Church to hear the address of Colonel Edward Sabine, president of the BAAS. The next morning, Tyndall attended the first meeting of section A (mathematics and physics), of which he was secretary. Before the meeting, he was introduced to the Roman Catholic bishop of Down and Connor, Dr. Cornelius Denvir, vice president of the section.[8] That contact was to prove instrumental to the success of Tyndall's first visit to Belfast. As he explained in his journal, Tyndall required the use of a galvanic battery to illustrate one of the two papers he was to give to section A. At the close of business, he and the bishop "took a car and drove off to [Denvir's] place." There they "got the components of the battery together," and Denvir "gave directions regarding [the] poles of [the] electromagnet."

Tyndall had arrived in Belfast as a lecturer in mathematics and natural philosophy at Queenwood's College, Hampshire. He was returning to Ireland after five years away with a strong sense of vocation and a "certain small celebrity." His aim was to "mingle among the magnates of the British Association," something he accomplished during the few days he was in Belfast. He participated fully in sectional business, speaking twice and taking part in the lively discussions after papers. He spoke for the first time to Sir David Brewster and Thomas Romney Robinson, and enjoyed a beer with the German physicist Emil du Bois-Reymond. Throughout his time in Belfast, he was chiefly concerned with consolidating his reputation as a scientific man and with evaluating other leading natural philosophers. His own record of his talks mentioned the enthusiastic response of his audience, both famous and obscure. These journal entries also displayed a fragile ego (something Tyndall admitted in a comment inserted some time later), Tyndall being somewhat paranoid about how others perceived him. He showed little concern with wider urban politics, religious or civic, in Belfast itself. The town was, in his own journal musings, largely a backdrop to the scientific and social proceedings of the BAAS.

It was Tyndall's encounter with the Catholic bishop that brought him closest to local religious and political concerns. Denvir, as Tyndall recorded, had "rendered him the greatest assistance" in setting up the experiments that illustrated his second talk to section A. The bishop repaid the compliment, commending Tyndall on his lecture. It was, Denvir remarked, "the best thing he had heard." The experiments elicited "eulogistic remarks" from William Thomson, and Denvir later encouraged Thomas Romney Robinson, director of the Armagh Observatory, to speak with Tyndall.[9]

This friendly collaboration with a bishop responsible for Belfast's growing Catholic population stands in stark contrast with Tyndall's dramatic verbal assault on the Catholic hierarchy in 1874 (and, with even more vitriol, in 1890). Equally, Denvir's obvious enthusiasm for Tyndall's science and his BAAS lecture was dramatically different from the virulent attacks featured in Belfast's Catholic newspapers printed during Tyndall's subsequent visits. The lack of polemic in 1852 is explicable in a number of ways. The most obvious is that Tyndall said nothing in his BAAS sectional talks that was likely to cause offense to those who disagreed with his views on religious matters. His subject matter was highly technical in nature. He spoke not as a celebrity or as president of the BAAS but as a young physicist keen to establish a scientific reputation. Equally, Tyndall's attitude toward religion left room for the kind of congenial relations expressed in his dealings with Denvir.

By 1852, Tyndall had moved some distance from his staunchly Protestant upbringing. As Geoffrey Cantor has shown, Tyndall had already warmly embraced the transcendental creed of Ralph Waldo Emerson and Thomas Carlyle, and was beginning to explore pantheistic conceptions of nature and the divine.[10] His attitude toward Catholicism was no longer defined by the vigorous polemics of Protestant reformers and Puritan divines. Tyndall had come to despise theological rigorism wherever found. So, while he was critical of Roman Catholicism as a religious institution and dogmatic system, he had no difficulty in admiring and praising individual Catholics who displayed either tolerance or an appetite for learning. Denvir, known for his liberal leanings, as was indicated by his support for the queen's colleges, fitted this mold.[11]

For Denvir, assisting Tyndall was one expression of a wider strategy that he had long adopted in relation to Belfast's fractured political and religious culture. While the number of Catholics in the town was growing, they nevertheless remained a minority and one without much influence. Denvir was sensitive to the strongly anti-Catholic sentiment that frequently erupted in Belfast and sought to appease such aggression through civic cooperation. He was widely criticized for being too timid in the face of Protestant pressure but held his ground, participating in causes such as nondenominational education that his more conservative fellow bishops increasingly decried.[12]

Denvir's assimilationist line and his cooperation with local and national government did not put him entirely out of kilter with his parishioners. As Sean Connolly has shown, Belfast at midcentury, while marked by sectarian tensions and intercommunal violence, also reflected some of the standard features of other burgeoning urban centers in Britain.[13] Cross-confessional efforts to bolster civic pride and display Belfast's progress as an industrial hothouse had marked the visit of Queen Victoria in 1849. Catholics, led by Denvir himself, were content to declare publicly their loyalty to the queen without suggesting it undermined their sense of Irish and Catholic cultural identity. Denvir's interest in science was central to his efforts to build good relations with Belfast's civic elite. His expertise in things electrical was quite literally put on show during civic events such as the annual Victoria fete held on Queen's Island, reclaimed land that later became the center of Belfast's expanding ship-building industry. More importantly, his scientific interests, cultivated while professor of natural philosophy at Maynooth, made possible his active membership of several local scientific and learned societies. These civic institutions facilitated links with other members of Belfast's elite and demonstrated goodwill toward Belfast's intellectual culture.

In these ways, Denvir was able to reinforce his reputation as a bridge-builder and trustworthy civic actor. Science provided a cultural matrix that allowed him to transcend local religious differences. His own formation as a Catholic natural philosopher had developed before he moved to Belfast. Its roots were in a classical education provided by the Presbyterian schoolmaster James Neilson in Downpatrick. Denvir's engagement with science also reflected a Catholic network of learning and research that promoted interest in electricity, most notably in the work of Luigi Galvani and Alessandro Volta. As professor of natural philosophy at Maynooth, Denvir had acted as mentor to Nicholas Callan, Denvir's successor, and, later, an internationally recognized expert in electromagnetism.

Denvir's interest in science was not unusual in early Victorian Belfast. The town boasted several societies promoting scientific knowledge among Belfast's educated classes. Such associational science was another feature Belfast shared with towns and cities across Britain, and, as was the case elsewhere, it connected the town with the wider world. The Belfast Natural History and Philosophical Society, of which Denvir was a member, was the most prominent and successful of the societies.[14] Among its chief objectives was the formation of a museum in which the citizens of Belfast would "behold the productions of the most distant and various countries," gathered by collectors who, for whatever reason, found themselves in such far-flung places. In the first decades of its existence, the society fostered an outward-looking, even cosmopolitan ethos and helped reinforce and celebrate Belfast's commercial, political, and religious links, formal and informal, with Britain's empire. The visit of the BAAS in 1852, in part made possible by the Belfast Natural History and Philosophical Society, provided an unprecedented opportunity to advertise the society's work. As with other civic societies, membership was dominated by Anglican and Presbyterian social elites.[15]

None of this is to suggest that Belfast in the first half of the nineteenth century was free from controversies provoked by scientific claims interpreted as antagonistic to revealed religion or political convictions. Twenty years previously, a particularly sharp dispute over the merit and implications of geological science had provided good copy for local newspapers over a period of four months. At its heart, the controversy involved a long-running public debate between John Edgar, a Presbyterian minister of the strictly orthodox Secession Synod, and Henry MacCormac, a prominent physician and heterodox social reformer. When MacCormac had hinted at a transformist account of life and at the "low antiquity" of humans in a lecture delivered under the auspices of the Belfast Academical Institute, Edgar had launched a bitter attack. The accusations hurled at MacCormac bore a strong family resemblance with those later directed at Tyndall by several of Belfast's Presbyterian clergy after his address of 1874. MacCormac, Edgar insisted, had overstepped the bounds of acceptable speech by insinuating opinions that went beyond the purview of a science lecture, particularly one sanctioned by the Belfast Academical Institute. Edgar was of the view that MacCormac had imported ideas under the guise of science that were politically explosive. He detected, for example, hints of Owenite socialism and a commitment to solutions to urban poverty that denied any role for divine assistance. It was, then, a contest over cultural authority as much as particular scientific claims, a much-discussed feature of the better-known controversy sparked by Tyndall in Belfast several decades later.[16] Twenty years later, when Tyndall spoke in Belfast for the first time, the potential for these kinds of disputes undoubtedly remained.

In 1852, the Belfast that Tyndall's conversations and communications pointed to was one that foregrounded civic cooperation and congenial civic relations between the town and science. This relatively scenic depiction emerged through informal conversation rather than from formal communication, the former being particularly important to a physicist on the make. Tyndall's brief accounts of the men of science that he met pointed to a rather different relationship among science, religion, and politics than the one evident in the earlier dispute over geology or in his accounts of later visits in 1874 and in 1890. In 1852, he generated no controversy. Instead, he elicited praise and received assistance from the leader of the town's small and poor but growing Catholic population. In ways that echoed Queen Victoria's positive portrayal of Belfast three years previously, Tyndall's impression was of a town able to support the pursuit of civic and intellectual progress and be open to the wider world.

A very different account of Belfast emerged when Tyndall returned to the town in 1874. To begin to grasp the immediate origins of that more negative portrait, the chapter moves forward to late 1872 and crosses the Atlantic to New York.

Re-placing Tyndall's Belfast Address

On Sunday morning, December 22, 1872, the Reverend John Hall entered the pulpit of Fifth Avenue Presbyterian Church in New York City. A celebrated preacher who was turning around the fortunes of his growing congregation, Hall took as his subject the topic of prayer. During the course of his sermon, he made mention of the visit to the city of John Tyndall. New York was awash with discussion about Tyndall's sensational lectures and, to the faithful, scandalous published views on the efficacy of prayer. Hall wished to combat Tyndall's skepticism about prayer and apparently did so in the most pointed manner. The content of Hall's sermon was not recorded, but Tyndall caught wind of the critique and responded in his next lecture, delivered in the hall of the Cooper Institute a few days later. At its close, Tyndall mentioned his regret that he had "aroused some bad feelings" in a number of Presbyterian clerics by "judging of matters by purely scientific rules."[17] This rather colorless aside was prompted by Tyndall's belief that Hall had accused him from his pulpit of lecturing in America simply to make money. That, to Tyndall, was the worst kind of insult. As it turned out, Tyndall was mistaken and issued a public half apology to Hall (or that "ill-tempered Irish Presbyterian," as Tyndall elsewhere described him) in the form of a letter to the *New York Times*.[18] Hall's nonconciliatory response appeared in the same paper the next day.[19] Tyndall, he argued, owed an apology to all the Presbyterians he had offended in his lecture and compared the celebrity physicist to a "third-rate politician."[20] The dispute rumbled on in the *Popular Science Review*, with Tyndall countering Hall's "intemperate" attack by accusing him of "wilfully twisting" his words.[21] On Tyndall's departure from New York, the controversy quickly blew over. As it turned out, however, the episode was one significant source of the much bigger storm that would hit Belfast when Tyndall spoke in the town's Ulster Hall as president of the BAAS less than two years later.

The exchange with Hall, which lingered in Tyndall's mind, reemerged in the wake of a meeting of the General Committee of the BAAS in Bradford on September 22, 1873. The main item of the meeting's agenda was the election of the president for the following year. The association was again to meet in Belfast, its first visit there since 1852. The discussion, according to the *Times*, was "warm," the heat coming when the mayor of Belfast, James Henderson, attempted to block the nomination of John Tyndall by arguing for a "local" candidate, Thomas Andrews, professor of chemistry at Queen's College Belfast.[22] As the discussion unfolded, Andrews's health ruled him out. But damage had been done. When Tyndall read the *Times*' report, his immediate reaction was "that the people in Belfast wished to have nothing to do with me."[23] As he rather ominously put it to Thomas Huxley: "They do not want me. Well in return I may be less tender in talking to them than I otherwise should have been."[24] Writing to his confidant Tom Hirst, he complained further that "they made it out, or rather an intemperate minister made it out in New York that I had attacked the Presbyterians, so that I was not surprised at a [denouncer] to my appearance among the Presbyterians of the North of Ireland."[25] The memory of Hall's insults had stayed with Tyndall, conditioning his reading of the *Times*' report. It took Huxley and William Spottiswoode to persuade him that Anderson's efforts were no more than an "asinine attempt" to privilege a local candidate.[26] Even so, Tyndall was clearly irked. It seems that the shadow of Hall played on Tyndall's mind when, "like Mahomet," he "retire[d] to the solitudes" to determine what he should say "to the people of Belfast."[27]

Tyndall delivered his presidential address in Belfast on the evening of August 19, 1874. Speaking in the Ulster Hall, surrounded by civic pomp and ceremony, Tyndall addressed an audience of nearly two thousand for an hour and forty-five minutes. Tracing the history of the growth of scientific inquiry from "primitive man" to his own time, Tyndall cast science as constantly threatened by overreaching religion. The heroes of his account resisted the forces of religious intolerance and investigated matter and life free from theological assumptions about its origins or operations. Tyndall insisted that religious sentiment was an "elemental bias in man" but argued that the form it took must not be fixed in creed or dogma. Kept within its own sphere of emotion and creativity, religious sentiment as an ancient motive force was vital to the success of science. The "domain of knowledge" was, however, strictly out of bounds, and any theological system that sought to control the advance of science and a scientific account of nature should be resisted to the death.[28]

In reaction, the pulpits of Belfast "thundered" at Tyndall, intensifying and multiplying Hall's original attack in New York.[29] This was not unexpected. Hall had been nurtured in the same religious environment as those who occupied Belfast's Presbyterian pulpits. He had trained in Belfast under the same Presbyterian mentors, including the Reverend Henry Cooke and the Reverend John Edgar.[30] The backlash against Tyndall emerged from a well-established tradition that balanced a commitment to divinely revealed knowledge with an exploration of nature, God's second book. Tyndall's co-option of induction as a method for investigating the natural world without religious interference or associations was particularly galling. Belfast's Presbyterians had built their understanding of the relations between natural and revealed knowledge on the premise that knowledge of nature uncovered by induction would not and could not conflict with divine revelation.[31] The dogmas of reformed Protestantism were protected by an inductive science that, in principle, could not contradict truths directly revealed in the Bible. For Belfast-based, and Belfast-trained, Presbyterians, this general formulation of the relations between theology and science was articulated in a way that narrowed the scope granted to scientific theorizing to a greater extent than other advocates of the same position. Despite, or perhaps because of, Tyndall's upbringing in a kindred Protestant tradition, he was repelled by this "Baconian compromise," especially as articulated and defended in Presbyterian Belfast (and New York).[32]

For all their fury toward an apostate from within their own staunchly Reformed Protestant tradition, Presbyterians were not alone in denouncing Tyndall in the pages of Belfast's newspapers. Local Catholic commentators used their own newspapers to take aim at an outright infidel dressed up in the ostensibly neutral dress of science. This local Catholic response has remained unreported in recent scholarly accounts of the Belfast reaction. On the other hand, official pronouncements of Ireland's Roman Catholic hierarchy against Tyndall have drawn some attention. As a number of scholars have suggested, one key motivating force in that elevated circle was Catholic control of education.[33] Tyndall presented the bishops with an opportunity for reinforcing their view that any education not overseen by the church could drift toward Tyndall-style materialism.[34] But the backlash among Belfast's Catholic community was differently styled and had other, more locally derived motivations. Certainly, it contrasted dramatically from the warm welcome offered to Tyndall by Cornelius Denvir in 1852. It also differed from the more distant condemnations issued by the bishops and did not follow the same track as the local Presbyterian reaction. In summary, hurling abuse at Tyndall provided a way to shore up Catholic interests in Belfast's civic economy.[35]

It was clear by 1874 that the kind of integration envisioned by Denvir during his period as bishop of Down and Connor had not materialized. Sharp and sometimes violent sectarian tensions remained. For the most part, Belfast's Catholics had kept themselves separate from the town's civic and educational institutions. The Queen's College, which had enjoyed the support of Denvir when it opened in 1849, attracted only a handful of Catholic students (against the advice of Denvir's successor, Patrick Dorrian) and included a solitary Catholic academic member of staff.[36] In 1858, the Belfast Catholic Institute was established to encourage the development of intellectual culture among the town's rapidly growing Catholic population. The institute quickly floundered when it became embroiled in politics, falling foul of Patrick Dorrian's censure on the back of accusations of harboring Fenianism on the one hand and, on the other, a less-than-respectful attitude toward ecclesial authority.[37] It was also subject to attack from Protestant and Unionist quarters. A columnist for the *Belfast Newsletter* got in early, making the accusation that the institute's lecture hall "will not merely exhibit the pleasing and instructive experiments by which young science delights to commend itself but may occasionally echo the sounds of Sepoyism and scarcely veiled sedition."[38]

It is not surprising, then, that when the BAAS met in Belfast in August 1874, there was little evidence of official Catholic involvement with its business. The meeting did attract close and sustained attention in two of the town's Catholic-owned newspapers, however. One of the central threads that wound its way through the voluminous commentary in the *Morning News* and, more especially, in the *Ulster Examiner* was the ongoing millworkers' strike.[39] The strike involved a dispute between a workforce that included large numbers of Catholics and exclusively Protestant (and mainly Presbyterian) masters. While Tyndall's supposed materialism and his antagonism toward theological infringement on ever-expanding scientific territory was criticized by Belfast's Catholic press, his presidential address quickly moved from being the object of religious critique to become an instrument useful for attacking local foes.

The opening salvo against the BAAS by the editor of the *Ulster Examiner* was aimed at the local organizers of its annual meeting. An editorial that appeared before the meeting started lambasted the extravagant spending on the weeklong congress. This, the editorialist judged, was in appalling contrast to the forty thousand millworkers who were being "starved out" by the town's Protestant industrialists. The strike over pay and conditions had already run for four weeks without resolution. According to the *Examiner*, the spectacle of "the sickly and pining child, the pale and emaciated mother" was kept from view as the visitors drawn to the town by the association's annual meeting started to arrive.[40]

This theme was repeated when the editor of the *Examiner*, Catholic priest Michael Cahill, offered an initial reaction to Tyndall's Ulster Hall address. Cahill's commentary was, in effect, a satire that played with the materialistic notes detected in Tyndall's speech. The priest-journalist asked his readers to consider whether Tyndall, staged as an advocate of the reduction of human society and economy to the movement of molecules, could concoct a "scheme so to direct the managing directors of the flax-spinning limited liability companies that the unfortunate molecule workers . . . could get twelve shillings-a-week instead of nine."[41] The answer was clear. Tyndall could not. The underlying message of Cahill's satire was that Tyndall's metaphysics supplied no basis for a just settlement to the crippling labor crisis that was destroying the lives of the town's millworkers.

In his next editorial, published a day later, Cahill continued his strategy of using Tyndall as an excuse to attack local figures unpopular with Catholics. His main target was the town's mayor, James Henderson, who had delivered a vote of thanks after Tyndall's address. In Cahill's telling, he had disgraced himself by telling poor jokes and dilating on the progress of Belfast's leading industries. The lack of concern Henderson displayed toward the "sickly and pining child, the pale and emaciated mother, the barefoot and poorly clad maiden"—or those "suffering atoms"—was consistent with Tyndall's demolition of "the great plan of creation."[42] Both undermined moral culture and championed materialism, whether philosophically or practically defined.

As Cahill continued his barrage of editorial commentary on Tyndall's address, he returned again and again to the juxtaposition of Tyndall's metaphysical materialism—which, Cahill believed, was apparent despite the physicist's strenuous efforts to qualify or deny it—and the economic and moral materialism of the oppressors of Belfast's millworkers. Cahill's animus against "Scotch colonists," who made Belfast's laborer's "toil without mercy for the merest pittance daily," was expressed alongside a further attack on the disastrous consequences of Tyndall's philosophy of nature. The latter dissolved the human intellect into the fortuitous motion of mindless matter, turning the BAAS into "a collection of learned baboons." The former had turned Belfast workers into dispensable cogs in a profit-maximizing machine indifferent to their humanity. The rapacious capitalism of "Scotch" millowners was consistent with Tyndall's amoral universe.[43]

While the plight of millworkers dominated Cahill's output, this was not the only quarry he pursued in taking Tyndall to task. He also used Tyndall's address to launch attacks on the Protestantism that dominated Belfast's political and civic life. If Tyndall's materialism and the capitalist agenda of Belfast's commercial class were closely connected, a line could also be drawn between the metaphysics touted by the BAAS's president and doctrines held dear by dedicated Protestants. This was particularly apparent in an editorial published by the *Examiner* a few days after the BAAS meeting had finished. Here Tyndall's atomism—the apparent reduction of life to the random motion of individual atoms—was compared to the Protestant view that "individual reason" trumped the "aggregate" authority of the church. Tyndall's "dicta" about atomic theory rang true in Protestant Belfast, where the conviction that "private judgement claims a dictatorship in theological reasoning" was widespread.[44] To the editorialist, the materialist metaphysics proclaimed by an apostle of science was the logical outworking of the elevation of individual interpretative whim over the weight of authoritative tradition. Protestantism was, at its roots, an incipient form of Tyndallism.

Several weeks later, the *Examiner* published another editorial dedicated to the furor over Tyndall, this time through a review of a sermon preached against Tyndall by a local Presbyterian minister, the Reverend John MacNaughtan. While the newspaper found MacNaughtan's sermon commendable in certain respects, the accusation that the Catholic Church had condemned Galileo's science was singled out as a serious flaw. Copernican science, the article suggested, had, in fact, been "respected and encouraged" by the Catholic Church. What had been justly condemned was the fact that Galileo had, by the lights of the time, "travelled outside the legitimate bounds of science."[45]

This was, the editorialist continued, precisely the same in the case of Tyndall. The BAAS's president had proclaimed unsubstantiated speculation as fact and built on it an anti-Christian metaphysics.

What is clear from these fulminations is the extent to which local urban and religious politics shaped the tone and substance of the debate. Local Catholic reaction was dominated by the complaint that Catholics were being systematically marginalized, economically and culturally, in Belfast. Tyndall provided the occasion to swing heavily against a "power geometry" that privileged Protestant interests and ideas.[46] Arguably, the editorials that appeared in the *Examiner* were, at base, much more exercised by local grievances than they were about either the content or the import of Tyndall's lecture. There was little reason for Belfast's Catholics to accord significance to the declarations of the president of an association that reinforced their marginal status. In contrast, there were many reasons to use the furor generated by Tyndall's address to air bitter complaints against millowners and a Protestant civic hegemony.

The local Presbyterian and Catholic reactions to Tyndall's address, when closely probed, reveal at least two things of note. First, Belfast's Presbyterian community extended well beyond the town's official limits. Tyndall's "less tender" tone was, at least in part, a reaction to his exchange with a Belfast-trained Irish Presbyterian in New York late in 1872. This indicates, among other things, that the Belfast "context" of Tyndall's address cannot simply be a matter of attending either to the immediate or the near at hand (whether in temporal or geographical terms). Second, and by the same token, acutely local labor disputes, largely ignored in other accounts of Tyndall's address, were crucial to the response of Catholic commentators in Belfast. The protests they made were not at the level of national politics but rather an intensely local grievance. Although the "local" here remains crucial to understanding the Belfast that emerged from, and helped to invest certain meanings in, Tyndall's address, it was given different expression by different groups and was connected to a more dispersed and diverse set of power relations in place.

Science for a New City: Tyndall, Belfast, and Home Rule

In late November 1889, the Belfast Society for the Extension of University Teaching held a meeting to decide on a speaker to give the inaugural address of a new series of lectures aimed at Belfast's educated public. The name settled upon was John Tyndall. Acting quickly, the president of the society, the

Reverend Thomas Hamilton, sent a letter to the eminent physicist. At first, Tyndall declined, citing his struggles with insomnia as the reason. Hamilton, however, persisted. He sent two more earnest letters, pleading with Tyndall to inaugurate the lecture series. In one he wrote that the success of the university extension lectures would be "assured by your presence, & our personal recollection of the pleasure which we derived from your last visit are so delightful that we cannot lightly give up the faintest hope of seeing you among us again." Noting that "the change of air" might prove a cure for insomnia, Hamilton added another reason for Tyndall to reconsider: "Pardon me for adding that your decided utterances on the question of the Union have endeared you more than ever to us here. Of course our new Society is non-political, but we know what we know."[47]

These last words were telling. In 1887, Tyndall had entered into a running debate with William Gladstone over Irish Home Rule. The exchange, which took the form of letters to the *Times*, represented Tyndall at his most polemical. He did not hold back from expressing his bitter opposition to Home Rule and to the former premier promoting it. On the back of Tyndall's efforts, the *Belfast News-Letter* portrayed him as a dazzling new opponent of Gladstone's Irish policy. Tyndall, "one of the greatest intellects this century has produced," was a despiser of alliances with "bigoted priests" and Ireland's separatist movement, and had wholeheartedly condemned the growing "fungus of Home Rule." What this scientific champion of the cause discerned was the appalling specter of Gladstonian opportunism and an Irish nationalism being fueled and manipulated by a benighted ultramontane Catholic clerisy. Tyndall provided the newspaper with plenty of rousing rhetoric to confirm his political credentials. Among them was Tyndall's declaration that he would sooner "shoulder [his] rifle among the Orangemen . . . than hand over the loyalists of Ireland to the tender mercies of the priests and Nationalists."[48]

Tyndall made efforts to offset his irascible rhetoric by using more measured language in his arguments against Home Rule. In an article published in *St James's Gazette*, Tyndall wrote of his empathy for the Young Irelanders, whose "songs . . . he knew . . . by heart." He remembered, too, the generosity another Roman Catholic bishop of Belfast, Patrick Dorrian, had extended to him a quarter of a century earlier.[49] But those wistful memories had to be set to one side. The alarm created by Gladstone's Home Rule campaign required the "good old spirit of Protestantism . . . the private judgement of brave men to confront the decrees of an immoral Pontificate."[50] The same appeal to a Protestant heritage was made in his final letter to the *Times* on August 9, 1887.

Ulster, Tyndall declared, "is strong enough to protect itself. The blood of the heroes of the Reformation still stirs its pulses." Tyndall worried, however, about the "loyal men" in the south of Ireland. It was among those men, threatened now by Home Rule, that Tyndall had "learnt to read and love my Bible." He could not stand aside and see them succumb to the rule of Ireland's "Romish priesthood," their "hereditary enemies."[51] It was not surprising that Tyndall's stirring words were quoted to applause during the opening of an Orange Hall in Ulster several months later.[52]

In all of this, Tyndall was attempting, as he later put it, to "out-Gladstone Gladstone," famous as the statesman was for the persuasive power of his political oratory.[53] But he had not yet taken to the political platform, using his oratorical skills honed over a long career of lecturing, to attack Gladstone's political program. The invitation to Belfast by the Reverend Thomas Hamilton presented him with an opportunity to do just that. As is clear from correspondence between Tyndall and Hamilton, there was a largely unspoken hope that Tyndall's visit to Belfast would strengthen the anti–Home Rule cause. Hamilton, whose appointment as president of Queen's College had been regarded by the Presbyterian Church in Ireland as an opportunity to reassert its influence over higher education in Belfast, was also strongly opposed to Gladstone's campaign.[54] As Tyndall's host in Belfast, he was the point of contact to facilitate Tyndall's involvement in local resistance.

The topic that Tyndall selected for his university extension lecture avoided any direct political or theological charge. The organizing committee asked him to speak on Pasteur, but Tyndall was not keen. He also rejected Carlyle as a topic, writing that it "will not do, for I fear, after what I have recently written regarding him I should have the dynamiters down upon me were I to lecture about him in Belfast."[55] In an article written for the *Fortnightly Review*, Tyndall had drawn attention to Carlyle's strong words opposing repeal of the Union, a move that would assuredly irk the Irish physical force nationalists then active in Britain.[56] Instead of Carlyle, and at the suggestion of his wife, Tyndall suggested a talk on microorganisms, "our invisible friends and foes." This had immediate relevance to a city ravaged by tuberculosis and influenza and was a subject that Tyndall was known to have investigated and spoken on many times. There was no need to worry that Tyndall would cause much offense to his Presbyterian hosts or to the capacity crowd that filled the Ulster Hall to greet him on the evening of January 21. The audience of twenty-five hundred had rushed to secure tickets to hear, as the mayor of Belfast put it, "one of the most brilliant popular expounders of science in the present day." In his vote of thanks after the lecture, Peter Redfern, professor of anatomy and physiology at Queen's College, praised Tyndall, that "master of exposition," for his powerful and delightful lecture. The Reverend R. J. Lynd, a prominent Orangeman and former moderator of the Presbyterian Church, warmly seconded Redfern's remarks. Tyndall was touched. He could not "translate into language the feelings that your kindness has raised in me."[57]

The warmth of the reception given to Tyndall's extension lecture was near universal. There were, however, a few dissenting voices. A ripple of discontent came from local temperance campaigners who took exception to Tyndall's comments about the curative properties of alcohol.[58] Their efforts failed to stimulate a wider outcry. A complaint also appeared in a letter written by the son of the celebrated local medical practitioner Henry MacCormac, leading disputant in the controversy over geology that had made local headlines in the 1830s. During his medical career, Henry MacCormac had tirelessly advocated fresh air as the key to preventing tuberculosis, which he regarded as caused by "pre-breathed air."[59] He had written against Robert Koch's experimental findings that pointed to a bacillary cause of the disease. His son, not a medical man, accused Tyndall of touting "invalid theories and empirical panaceas" and ignoring the "plain truth" that those who got "plenty of God's fresh air into their lungs . . . need fear no consumption."[60] There was a certain irony to this. MacCormac had decades earlier defended an old earth geology and a quasi-naturalistic science against bullish Presbyterian detractors. Now, his son was defending his father's science of fresh air against Tyndall while more orthodox Presbyterians lined up to praise the physicist's scientific reputation.

There is little doubt that MacCormac's ideas were considered dated and defunct. For those still active in promoting public health in Belfast, Tyndall's address was a boon, useful for helping a struggling charitable institution, the Belfast Hospital for Consumption and Diseases of the Chest. The hospital's annual meeting took place the day after Tyndall's address, and the famous lecturer was several times mentioned to lend support to the institution. The plans to develop the hospital were based, in part, on the contagious nature of tuberculosis: Tyndall's authority was used to secure that crucial claim. Both the mayor and the president of Queen's College backed the premise and the development plans, thus lending civic weight to Tyndall's scientific pronouncements.[61]

The apparently innocuous nature of Tyndall's topic nevertheless allowed him to repeat in moderate language a point he had made in more polemical style sixteen years previously. In 1874, he had notoriously declared that "all religious theories, schemes and systems which embrace notions of cosmogony . . . must submit to the control of science." He was then in such earnest that he went as far as to urge that any theological edifice that exercised "despotic sway over the intellect" should be "opposed, to the death if necessary."[62] This rhetoric of violence was not repeated in his university extension lecture. In that calmer discourse, he reminded his audience that the "only authority in science is the voice of Nature, interpreted by sound reason, or answering to well devised experiment." For all that, there were hints that his penchant for polemics was only dormant. In his stirring peroration, he ominously warned his audience that there were still "those among us ready to oppose the plainest dictates of reason."[63]

After Tyndall's lecture, he found himself in receipt of invitations to meet both conservative and liberal groups concerned about Home Rule. One invitation addressed to Tyndall was by the Reverend Robert Kane, grand master of the Orange Order in Belfast and Church of Ireland rector of Christ Church, Belfast. Kane's message, which represented the wishes of the Belfast Conservative Association, urged Tyndall to speak to a "Unionist Demonstration" organized to celebrate the contributions of Lord Londonderry, the lord lieutenant of Ireland, to the Unionist cause. Tyndall would move the meeting's first resolution by "offering our warmest congratulations to her Majesty's Ministers on the signal success which has crowned their Irish policy." Tyndall was being called upon to give a vote of confidence in Salisbury's anti–Home Rule government. After seeking counsel, including from Thomas Hamilton, Tyndall agreed.[64] To prepare, Tyndall and his wife, Louisa, traveled to Newcastle, County Down. There at the Annesley Arms Hotel in the shadow of Slieve Donard, Tyndall's anti–Home Rule screed was put to paper.

The "Unionist demonstration," held on January 28, drew a large crowd to the Ulster Hall. The *Belfast Newsletter* reported a packed auditorium. When Lord Londonderry came forward to address the loyal citizens of Belfast, the crowd rose "*en masse* waving their hats and their handkerchiefs." Once the lord lieutenant had finished a long and rousing address, the chair of the meeting, the Duke of Abercorn (cousin of Tyndall's wife), introduced the famous physicist to "loud and prolonged cheering." Tyndall, he declared, was "a friend" and "not unknown in Belfast." He had "at great trouble and considerable time . . . done his utmost to promote the cause of the Union." Stepping forward to deliver the meeting's first resolution, Tyndall delivered a long and vitriolic discourse against the threat of "unrighteous legislation" and spent much of his speech condemning Gladstone, "the wickedest man of our day and generation." He used his friendship with Charles Darwin to underscore a general principle that, in Tyndall's judgment, Gladstone had never followed: "the necessity in all true work of combining reflection in a high ratio with expression." Shifting tack toward the end of his address, Tyndall appealed to his hearers' civic pride: "My lords and gentlemen, I have walked through this city and around its suburbs with my eyes open. I was here some sixteen years ago and can mark the progress you have made since that time. In this respect I would back Belfast against any city in the United Kingdom. (Cheers.) You have your docks, your factories and your hospitals. New and noble public buildings have arisen. New streets which rival, if they do not surpass, the best streets of the English metropolis." Having evoked what might be lost if Home Rule became a reality, Tyndall concluded his speech with a call to the loyalists of Ulster to defend their "rights and liberties with the sword." He added that if Gladstone got his way, "tens of thousands of British men would be ready to leap into [their] ranks and to help [them] overthrow their foes."

Sir Edward Harland, the member of Parliament for North Belfast and celebrated shipbuilder, seconded Tyndall's resolution. Praising Tyndall as "a fellow countryman of whom I am truly proud," Harland pronounced him "one of the greatest living men enjoying a universal reputation." His "unbiased, unprejudiced view" of the Irish question would have an "immense effect with the deep-thinking men in England and Scotland."[65] Like Tyndall, Harland also dilated on the industrial progress of Belfast, showcasing the city as evidence of the tremendous benefits of current constitutional arrangements. Rule from Westminster underwrote Belfast's wealth and progress, and there was no reason why other urban centers in Ireland could not follow the same course. Harland was, however, circumspect, avoiding direct criticisms of public figures. Tyndall, he averred, had said enough to make clear who the enemies of Belfast's prosperity were.

The impression Tyndall and Harland gave was one of a booming industrial city proud of its success and anxious to display that progress through monumental architecture and other material expressions of civic pride. Belfast had had city status granted in 1888, testimony to its remarkable growth. The transformation of the urban environment in and around the city since Tyndall had last been there in 1874 was dramatic. Buildings such as the Central Library, the six-story Robinson and Cleavers department store, and the Ulster Reform Club were prominent indicators of intellectual, economic, and social progress.[66] What this kind of transformation did not communicate was the parallel increase in violence on Belfast's streets. In 1886, two years before Belfast celebrated city status, disturbances associated with the First Home Rule Bill led to an unprecedented thirty-two deaths. This reflected a trend of increased politically charged violence and killing in Belfast in the later decades of the nineteenth century.[67] While Tyndall did not discuss the real violence associated with debates over Home Rule, he continued, while in Belfast, to lend support to those opposed to it. The day after his speech at the demonstration, Tyndall dined with members of Belfast's Liberal Unionist Association, the other political wing of anti–Home Rule in Belfast.

Tyndall's own impression on leaving Belfast was that his visit had been a success. As he wrote to Tom Hirst, "the visit of your old friend to the place in which he gave such offence sixteen years ago, [seems more like] a splendid dream to me than anything else."[68] Certainly, there had been no reports of Presbyterian ministers using their pulpits to condemn the appearance of the famous "materialist" in the Ulster Hall, this time not once but twice. Yet it would be a mistake to think that his interventions in Belfast left no impression on the city or on those interested in the tense political situation there.

Some commentators enthusiastically affirmed Tyndall's impressions and trumpeted the many benefits of hosting the scientific luminary, not least the political capital gained from giving him a platform in Belfast. An editorial in the *Belfast Newsletter* was especially ebullient in its praise. The day after the demonstration, the editorialist quoted with relish the "sword" passage from Tyndall's speech and, in an echo of Tyndall's university extension lecture, warned readers to "take care" not to increase the strength of their "invisible foes."[69] This direct echo of the title of Tyndall's extension lecture was surely not lost on readers. In his scientific address, Tyndall had revealed to his listeners the usually invisible world of microorganisms. At its close, he had hinted that the advance of science was still under threat by opponents of reason. In his anti–Home Rule speech, he identified them: "Archbishop Croke and his myrmidons, backed by the ignorant and excitable peasantry of the south." The "noble Queen's College" that Tyndall praised as "ministering" to Belfast's intellectual needs would be swept away if the alliance of Croke and Charles Stuart Parnell—the political champion of Home Rule in Ireland—succeeded.[70] As well as supporting the agitation against landowners, Croke had been instrumental in making Parnell's parliamentary party the political voice for defending Catholic interests in the education question, interests that to Tyndall were inimical to the freedom necessary for scientific inquiry. There was no question about who Tyndall thought were the enemies of science and, by extension, humanity. The "political priests" agitating for Home Rule, now fully exposed by Tyndall, were a foe more deadly than the bacillus of anthrax.

In light of such partisan goading, it is not surprising to find the response of the nationalist press to Tyndall's political speech-making in the Ulster Hall was one of outrage and mockery. Tyndall's "sword" speech, as the Dublin-based *Freeman's Journal* dubbed it, confirmed to nationalists in Belfast and beyond that the celebrity scientist had degenerated into a rabid anti–Home Ruler.[71] The local Catholic press in Belfast was quick to point out that Presbyterian Unionists of various stripes were courting the very man that only sixteen years before had been roundly condemned for his godless and materialist manifesto. The *Belfast Morning News*, commenting on Tyndall's anti–Home Rule diatribe, excoriated him for crossing the line between science and politics and found his "politico-scientific speech" filled with "blatant balderdash and ill-concealed vanity." The paper also reminded its readers that when Tyndall had last been in Belfast, he had "flung the epithet of pigmies at the whole body of the Presbyterian Clergymen."[72] It was, in fact, Thomas Henry Huxley who had "flung the epithet," but the general point was clear. Presbyterian Belfast was now feting someone who a few years earlier they had pitilessly vilified.

In 1890, Tyndall was being depicted by the nationalist press as a representative—an archetype—of a "class" of Irish citizens who had for generations benefited from the kind of patronage that had left the vast majority of Catholics without adequate political representation or educational opportunities. An article in the *Nation* published shortly after his appearance at the Unionist demonstration in Belfast stressed exactly this point. Tyndallism, it declared, "was suckled on the corruptions of Carlow Orangeism and rocked in the cradle of Orange patronage." The Irish-born physicist, it added, wished to preserve the obscene privileges of his youth and to keep "the road to Tubingen" open only for those who subscribed to his "landlordism."[73] In other words, his education and success in science—Tübingen is where Tyndall had trained as a chemist—was a clear sign of privilege propped up by a counterbalancing Catholic and nationalist disadvantage.

It was not, however, only the nationalist press in Ireland that took aim at Tyndall's political intervention. Although his speech was not given verbatim beyond Belfast, its general tenor and some of its more pointed remarks generated vigorous reactions. One of the first to respond was William Gladstone, who wrote to Tyndall to complain that, in his Belfast speech, he had put words in his mouth. Gladstone, Tyndall had declared, had "waited until he was 76 years old to discover that [William] Pitt was a blackguard." Gladstone denied describing the architect of the Act of Union in that way and asked Tyndall to provide his source.[74] A few days later, the prominent Liberal MP William Harcourt came to Gladstone's defense, writing to the *Times* that Tyndall's "disgraceful" attack was a perfect example of "how little education and science are able to restrain men of violent and intemperate passions."[75] In a later letter to the *Times*, Harcourt again complained about the "unphilosophical violence of [Tyndall's] Ulster oration" and used Francis Bacon's words to accuse Tyndall of being overcome, against a scientific spirit, by "the humours of the affection." It had been unwise of Tyndall, a "scientific Orangeman" to imitate the "spirit and manners" of Orangeism, "especially in Belfast."[76]

Tyndall had certainly given his critics ammunition. There surely was something incongruous about the glowing reports in Protestant papers of Tyndall's science lecture and political speech. While it was true that Presbyterian clergymen of a Liberal Unionist stripe did not attend the demonstration, it was political more than religious scruples that kept them away. Tensions remained between the conservative and liberal wings of Belfast's anti–Home Rule movement. For all that, the impression given to the *News-letter* that "Tyndallism"—or the affirmation of Tyndall's brand of evolutionary naturalism—was relatively uncontroversial by 1890 among all but a tiny minority was almost certainly false.

Tyndall's appearances in Belfast in 1890, rather than generate a "winter of discontent," as he had in 1874, took advantage of one. The Unionist demonstration confirmed the rising tide of political inquietude in Belfast. Tyndall's cooperation with Protestant Unionists of different stripes also built upon and sharpened an alliance that he had formed ever since the first Home Rule Bill was proposed and defeated several years earlier. He continued to play up the glories and inherited liberties of Protestantism while in Belfast. To complain now that Tyndall the arch-materialist was, again, given a platform in the Ulster Hall would have been a strategic blunder by even the most religiously conservative anti–Home Ruler. In identifying the enemies of reason and intellectual freedom, Tyndall narrowed his target. Roman Catholicism, politicized by priests and their leaders, was the real foe.

Dogmatic Protestantism no longer seemed a threat to Tyndall's vision of a world free from the tyranny of overextended theological systems. Protestant Belfast, as he observed, was booming—and not just commercially. Its hospitals and educational institutions were evidence of monumental progress.

That its public health officials were formulating plans based on Koch and Pasteur spoke of the triumph of scientific reason. The Queen's College, led by the enlightened Presbyterian Thomas Hamilton, was providing nonsectarian education to an energetic and ambitious Protestant population. Belfast was a jewel in the crown of the British Empire and a key defense against threats to Britain's imperial project. Tyndall could not have propagated this version of the city on either of his previous two visits. The transformation in the physicist's view of Belfast was remarkable.

Conclusion

When John Tyndall first visited Belfast to speak as a young man of science in 1852, what stood out was cooperation with a friendly and tolerant Roman Catholic bishop. Here was a town that had attracted a Catholic clergyman of a tolerant and liberal stripe who was not only friendly toward science but also an expert in the very area of research that Tyndall was most interested in. The uneasy equipoise that characterized some accounts of Belfast in the early 1850s was one that Tyndall experienced firsthand and shared with others. It was a Belfast that sat well with Tyndall's own priorities during his first visit. He worked hard to cooperate and collaborate in ways that aided his own growing reputation as an ambitious young physicist. For Tyndall in 1852, there was no place in Belfast for political or metaphysical disputation.

His second visit was dramatically different. Quite deliberately, Tyndall launched a rhetorical attack on the opponents of science and intellectual freedom. As Tyndall saw it, Belfast was full of Presbyterians of the type that had annoyed him so much in New York two years previously. They would, he convinced himself, be resolutely set against him and his vision of theology-free science and the liberated society that he believed could be built upon it. With Denvir no longer in post and the Catholic Church working against efforts to teach science without religious oversight, Tyndall also made sure his Belfast address cast a plague on Roman Catholicism. The Presbyterians reacted strongly and, a few months later, launched a full-blown lecture series to protest against Tyndall's metaphysical pronouncements. Belfast's Roman Catholic commentators also poured scorn on Tyndall but quickly moved on to use his address to target local millowners, Protestant civic leaders, and apologists for Presbyterianism. This was no longer a town marked by civic as well as scientific cooperation. Instead, a scientific address became a battleground that reflected and refracted a town torn apart by sectarian rancor and rivalry.

In 1890, Belfast looked different again. In contrast to 1852, Tyndall now forged an alliance with Belfast's enlightened and loyal Presbyterians. Catholics alone had become the confirmed and deadly enemies of reason and liberty: Protestants were the drivers of progress and patriotism. Tyndall, revealer of nature's secrets, also called out the invisible foes of Britain's empire and Ireland's place within it. The contagion of political priestcraft had to be contained and condemned. It was fortunate, then, that just as Belfast's hospitals, armed with the findings of Pasteurian science, were battling against bacilli, its citizens, with the help of Tyndall, were fighting the deadly infection of Home Rule. Belfast had become, to Tyndall, a shining beacon in the struggle against the dark irrationalism of Gladstonian politics then sweeping Britain.

Tyndall's various visits operated with and within three different Belfasts. Asking whether these tally with, or distort, some comprehensive urban reality has not been a key concern here. What Belfast was at different times cannot be exhaustively captured by any single description. Nevertheless, the Belfast in which the activities of Tyndall and his local allies and antagonists and critics occurred was to them real enough. It was the matrix in which Tyndall's own activities, and the local reactions to them, took form. At the same time, it was not a stable or entirely pregiven reality that influenced events in any straightforwardly mechanical way. It was enunciated and enacted in and through the actions and reactions associated with Tyndall's visits without being reducible to those variously situated performances. Tyndall and those who interacted with him certainly helped to script—sometimes literally when they stood up to speak—a particular but plural kind of place called Belfast.

These reflections point to some more general considerations related to the historical geographies of science communication. The interplay between the life spaces of an influential spokesperson and the dynamic and contested geographies of an urban culture was productive of, and positioned within, a complex and crosscutting set of relations and political realities. The town (then city) of Belfast never named a singular kind of place that provided a stable backdrop to a set of communicative acts. Nor can the place where Tyndall spoke be defined in terms of a particular scale of analysis. Tyndall's appeals to the macroscopic "wider town" and the microscopic "bacillus of anthrax" echoed the multiscalar character of his platform pronouncements. From the performing body of the speaker, through the mediation of print and institutional practices, to the transnational travels of the published speeches, those pronouncements, and the responses they provoked, resist being categorized using the conventional categories of scale. A case can be made for Catholic and Presbyterian engagements with Tyndall as "local" in nature, at least as a first approximation. But they were never local *simpliciter.* In the same way, Tyndall's rhetoric reverberated well beyond Belfast but never in ways that were independent from the particularities of the other places where it was reproduced and read. Attending to the multiscalar character of the communication and public contestation of science underlines the importance of careful attention to the relational and dynamic nature of the places where communication and contestation occur.

Notes

1. David N. Livingstone, *Dealing with Darwin: Place, Politics, and Rhetoric in Religious Engagements with Evolution* (Baltimore: Johns Hopkins University Press, 2014), 8.

2. Doreen Massey, *Power Geometries and the Politics of Space-Time* (Heidelberg: Hettner Lecture, Institute of Geography, University of Heidelberg, 1999), 22.

3. For example, John A. Agnew, "Space and Place," in *Handbook of Geographical Knowledge*, ed. John A. Agnew and David N. Livingstone (London: Sage, 2011), 316–30; Charles W. J. Withers, "Place and the 'Spatial Turn' in Geography and in History," *Journal of the History of Ideas* 70 (2009): 637–58.

4. For a nominalist challenge to ascribing agency to place, see Robert J. Mayhew, "Historical Geography, 2009–2010: Geohistoriography, the Forgotten Braudel and the Place of Nominalism," *Progress in Human Geography* 35 (2011): 409–21.

5. David N. Livingstone, *Putting Science in Its Place: Geographies of Scientific Knowledge* (Chicago: University of Chicago Press, 2003), 7.

6. Livingstone, *Dealing with Darwin*, 58–88.

7. For example, Bernard Lightman, "Scientists as Materialists in the Periodical Press: Tyndall's Belfast Address," in *Science Serialized: Representations of the Sciences in Nineteenth-Century Periodicals*, ed. Geoffrey Cantor and Sally Shuttleworth (Cambridge, MA: MIT Press, 2004), 199–237.

8. Denvir served as Roman Catholic Bishop of Down and Connor between 1835 and 1865 and died on July 10, 1866.

9. All the consecutive quotes in preceding paragraphs are from Tyndall Journal, September 2, 1852, Royal Institution (hereafter RI) MS JT2/V/p. 147.

10. Geoffrey Cantor, "John Tyndall's Religion: A Fragment," *Notes and Records of the Royal Society* 69 (2015): 419–36.

11. On Denvir's place among Irish bishops, see Ambrose MacAuley, *Patrick Dorrian, Bishop of Down and Connor* (Dublin: Irish Academic Press, 1987), 26–64 and 87–112, and Sean J. Connolly, "Paul Cullen's Other Capital: Belfast and the Devotional Revolution," in *Cardinal Paul Cullen and His World*, ed. Dáire Keogh and Albert McDonnell (Dublin: Four Courts Press, 2011), 289–307.

12. John J. Silke, "Cornelius Denvir and the 'Spirit of Fear,'" *Irish Theological Quarterly* 53 (1987), 130–43.

13. Sean J. Connolly, "Like an Old Cathedral City: Belfast Welcomes Queen Victoria, August 1849," *Urban History* 39 (2012): 571–89.

14. Ruth Bayles, "Understanding Local Science: The Belfast Natural History Society in the Mid-Nineteenth Century," in *Science and Irish Culture*, ed. David Attis and Charles Mollan (Dublin: Dublin Royal Society, 2004), 139–69. Jonathan J. Wright, "'A Depot for the Productions of the Four Quarters of the Globe': Empire, Collecting and the Belfast Museum," in *Spaces of Global Knowledge: Exhibition, Encounter and Exchange in an Age of Empire*, ed. Diarmid A. Finnegan and Jonathan Jeffrey Wright (London: Routledge, 2015), 143–66.

15. On the middle-class culture in Belfast at this time, see A. Johnson, "The Civic Elite of Mid-Nineteenth-Century Belfast," *Irish Economic and Social History* 43 (2016): 62–84.

16. For a full account of this episode, see Jonathan Jeffrey Wright and Diarmid A. Finnegan, "Rocks, Skulls and Materialism: Geology and Phrenology in Late Georgian Belfast," *Notes and Records of the Royal Society* 72 (2017): 25–55.

17. Anon, "Tyndall's Fourth Lecture," *New York Times*, December 27, 1872.

18. Edward C. Youmans, "A Correction: Letter from Prof. Tyndall," *Popular Science Monthly* 3 (1873): 243.

19. Anon, "Prof. Tyndall and the Presbyterians," *New York Times*, December 30, 1872. That it was Hall is confirmed in Thomas C. Hall, *John Hall: Pastor and Preacher* (New York: Fleming H. Revell Company, 1901), 275.

20. Anon, "Letters to the Editor," *New York Times*, December 31, 1872.

21. Youmans, "A Correction," 243.

22. Anon, "The British Association," *Times*, September 23, 1873, 5.

23. John Tyndall to Thomas Archer Hirst, September 27, 1873, RI MS JT/1/T/707.

24. John Tyndall to Thomas Henry Huxley, September 24, 1873, RI MS JT/1/TYP/9/3022.

25. Tyndall to Hirst, September 27, 1873.

26. Tyndall to Hirst, September 27, 1873.

27. John Tyndall to F. W. Farrar, March 17, 1874, Wellcome Library MS. 7777/90.

28. John Tyndall, "Inaugural Address," *Belfast Newsletter*, August 20, 1874, 6–7.

29. "Every pulpit in Belfast thundered at me": John Tyndall to Thomas Archer Hirst, August 26, 1874, RI MS JT/1/T/715.

30. Andrew Holmes, "Presbyterians and Science in the North of Ireland before 1874," *British Journal for the History of Science* 41 (2008): 541–65.

31. Holmes, "Presbyterians and Science in the North of Ireland before 1874."

32. On the Baconian compromise, see James R. Moore, "Geologists and Interpreters of Genesis in the Nineteenth Century," in *God and Nature: Historical Essays on the Encounter between Christianity and Science*, ed. David C. Lindberg and Ron. L. Numbers (Berkeley: University of California Press, 1986), 322–50.

33. Livingstone, *Dealing with Darwin*, 74–76; Don O'Leary, *Irish Catholicism and Science* (Cork: Cork University Press, 2012), 28–32.

34. On the pastoral letter, see O'Leary, *Irish Catholicism and Science*, 28–32. The full text appeared, among other places, in *Freeman's Journal*, October 31, 1874.

35. For a full account, see Diarmid A. Finnegan and Jonathan Jeffrey Wright,

"Catholics, Science and Civic Culture in Victorian Belfast," *British Journal for the History of Science* 48 (2015): 261–87.

36. T. W. Moody and J. C. Beckett, *Queen's Belfast 1845–1949: The History of a University*, 2 vols. (London: Faber, 1959).

37. On the Institute and its demise, see MacAuley, *Patrick Dorrian*, 140–52.

38. Anon, *Belfast Newsletter*, October 12, 1858.

39. See, for example, *Belfast Morning News*, August 20 and 21, 1874.

40. Anon, *Ulster Examiner*, August 15, 1874.

41. Anon, *Ulster Examiner*, August 20, 1874.

42. Anon, *Ulster Examiner*, August 21, 1874.

43. Anon, *Ulster Examiner*, August 24, 1874.

44. Anon, *Ulster Examiner*, August 29, 1874.

45. Anon, *Ulster Examiner*, September 16, 1874.

46. Massey, *Power Geometries*.

47. Thomas Hamilton to John Tyndall, December 9, 1889, JT/1/H/14.

48. Anon, *Belfast Newsletter*, June 8, 1887, 5.

49. John Tyndall, "Reminiscences and Reflections," *St James Gazette*, April 23, 1887, 5.

50. John Tyndall, *Mr. Gladstone and Home Rule* (Edinburgh: William Blackwood and Sons, 1887), 4–5.

51. John Tyndall, "Professor Tyndall on Mr. Gladstone," *Times*, August 9, 1887, 4.

52. Anon, "Opening of Gortfad Orange Hall," *Belfast Newsletter*, November 5, 1887, 7.

53. John Tyndall to Charles Grant, February 8, 1889, RI MS JT/1/T/454.

54. Moody and Beckett, *Queen's Belfast*, 324.

55. John Tyndall to Thomas Hamilton, December 22, 1889, RI MS JT/1/T/467/2.

56. See John Tyndall, "Personal Recollections of Thomas Carlyle," *Fortnightly Review* 47 (1890): 5–32. On physical force Irish nationalists, see Niall Whelehan, *The Dynamiters: Irish Nationalism and Political Violence in the Wider World, 1867–1900* (Cambridge: Cambridge University Press, 2012).

57. Anon, "Our Invisible Friends and Foes," *Belfast Newsletter*, January 22, 1890, 7.

58. J. B. Wylie, "Dr Tyndall and Alcohol," *Belfast Newsletter*, February 4, 1890, 7.

59. Helen Andrews, "Henry MacCormac," in *Dictionary of Irish Biography*, ed. James McGuire and James Quinn (Cambridge: Cambridge University Press, 2009).

60. John MacCormac, "Dr Tyndall and Consumption," *Belfast Newsletter*, January 24, 1890, 7.

61. Anon, "Hospital for Consumption and Diseases of the Chest," *Belfast Newsletter*, January 24, 1890, 7.

62. Anon, "The Meeting in the Ulster Hall," *Belfast Newsletter*, August 20, 1874, 7.

63. Anon, "Our Invisible Friends and Foes," 7.

64. John Tyndall to Thomas Hirst, January 31, 1890, RI MS JT/1/T/819.

65. Anon, "Great Unionist Demonstration in Belfast," *Belfast Newsletter*, January 29, 1890, 6.

66. On Belfast's development through this period, see Stephen Royle, *Portrait of an Industrial City: Clanging Belfast, 1750–1914* (Belfast: Belfast Natural History and Philosophical Society, 2011).

67. S. J. Connolly and Gillian McIntosh, "Whose City? Belonging and Exclusion in the Nineteenth-Century Urban World," in *Belfast 400: People, Place, and History*, ed. S. J. Connolly (Liverpool: Liverpool University Press, 2012), 237–70.

68. Tyndall to Hirst, January 31, 1890.

69. Anon, *Belfast Newsletter,* January 29, 1890, 4.

70. Anon, "Great Unionist Demonstration," 6.

71. Anon, "Lord Londonderry in Belfast," *Freeman's Journal*, January 29, 1890, 7.

72. Anon, *Belfast Morning News*, January 30, 1890.

73. Anon, "A Scientific Orangeman," *The Nation*, March 15, 1890, 9.

74. Gladstone's letter is published in Anon, "Professor Tyndall and Mr. Gladstone," *The Times*, March 10, 1890, 4.

75. Anon, "Sir William Harcourt and Professor Tyndall," *The Times*, February 11, 1890, 10.

76. Anon, "Professor Tyndall and Sir William Harcourt," *The Times*, February 14, 1890, 12.

PART TWO

NATIONAL STUDIES

CHAPTER THREE

Henry Hotze in Place

Religion, Science, Confederate Propaganda, and Race

MARK NOLL

In Henry Hotze's remarkable career as a defender of slavery, publicist for the Confederate States of America, and apostle of racial inequality, place mattered. When, on January 1, 1856, Hotze dispatched a defensive letter to Arthur Comte de Gobineau in Paris, he wrote as a translator trying to explain why the author's book needed to be so substantially abridged and so extensively introduced for the translation he published in the United States: "An argumentative subject like yours, must be treated here somewhat differently from what it can be in France."[1] Seven years later, writing from London, Hotze explicated essential differences of European national character as he urged the Confederate secretary of state to keep sending the funds he badly needed to sustain his newspaper, the *Index*, which was the South's main propaganda organ in the United Kingdom. As he wrote on September 26, 1863, to Judah Benjamin: "It is much easier for the English, accustomed to a hierarchy of classes at home and to a haughty dominion abroad, to understand a hierarchy of races, than it is for the French[,] the apostles of universal equality."[2] Hotze thereby underscored how much the realities of geography intersected with the exigencies of nation-states. Yet from before 1856 and well past 1863, the structure of Hotze's principled convictions did not change. He remained, to at least some degree, a professing Christian with at least some respect for the Christian Scriptures. His commitments to the Southern system of American slavery and, after war broke out, to the Confederate States of America were much stronger. Even stronger still remained his belief in the ineradicable character of hierarchical racial distinctions. If, however, Hotze's priorities stayed the same, his advocacy for the causes he treasured differed significantly depending on where he was and to whom he was speaking. It would be difficult to find a historical figure whose career so thoroughly engaged the subjects (polygenesis, preadamitism, evolution) that have been illuminated by David Livingstone and, at the same time, so vividly illustrated the necessity of putting scientific disputes in their place, for which Livingstone has also advocated so persuasively. As with Livingstone's case study of the reception of Darwinism in South Africa, so Hotze's peripatetic career "demonstrates the continuing need to resist the inclination to trade in philosophical generality by abstracting the impact of science from the particularities of period and place; instead it invites us to further purse the project of uncovering the diverse historical geographies of science by probing local encounters with new ideas and tracing how they were mobilized for interests of very different kinds."[3]

First Nott, then Gobineau

Henry Hotze was a twenty-year-old immigrant from Switzerland working as a private tutor near Mobile, Alabama, when, in late 1854, one of his new American friends, Josiah Clark Nott, approached him with a business proposition.[4] Nott had only just finished work with his collaborator, George Gliddon, on a blockbuster book published earlier the same year. Their *Types of Mankind: Or, Ethnological Researches Based upon . . . Crania of Races, and upon Their Natural, Geographical, Philological, and Biblical History* synthesized the craniological research of a late Philadelphia physician, Samuel George Morton, the theories of Harvard zoologist Louis Agassiz about the geographical distribution of different human species, Gliddon's on-the-ground investigations of ancient Egyptian skeletons, Nott's reading of books on ancient history and natural science, and the long-standing disdain of Nott and Gliddon for orthodox theologians ("skunks" was their word) who tried to harmonize contemporary ethnology with biblical accounts of human origins.[5] The book's eight hundred pages, its prolific quotations from contemporary researchers, its many maps, and its even more numerous sketches of skulls and faces demonstrated, according to the book's preface, that "the diversity of races must be accepted by Science as a *fact*, independently of theology." In making this universal claim, Nott and Gliddon felt compelled to rebut universal claims of theology and also to acknowledge their American context. For the first, *Types of Mankind* continued the assault on scriptural science they had carried on for many years; in his section, for example, Gliddon held that "commentaries of the genuine English evangelical school" on the Table of Nations found in Genesis 10 were "of trivial value in themselves" and carried "less weight in science." For the second, American controversies over slavery undergirded the book's entire effort. As an instance, it took special care to show that even when Africans were brought to the United States as slaves, they remained physically and intellectually distinct from Caucasians, except where black-white "illegitimate consequences" had "deteriorated the white elements in direct proportion that they are said to have *improved* the blacks."[6]

This much-noticed volume immediately became the signature statement of the American School of Ethnology. Its tenets were crystal clear: humankind was divided into separate species (polygenesis), those species ranked themselves from inferior (black) to superior (white), conclusions concerning these distinct species came completely from careful scientific investigation, and Scripture was either irrelevant or flat-out worthless for serious ethnological research.

As might be expected, *Types of Mankind* generated immense controversy. It particularly offended the Reverend John Bachman, a Lutheran clergyman from Charleston, South Carolina, who had earlier engaged both Morton and Nott ardently to defend both scriptural accounts of human origins and the unity of the human race (monogenesis). Yet to intimate the importance of geography for assessments of all such debates, Bachman, whose family owned slaves, supplemented his ardent biblical monogenism with a comparably ardent defense of the American slave system.[7] In Europe, a distinguished readership also pored over *Types of Mankind* along with criticisms of the book, including Bachman's. They did so, however, with *their* local circumstances in view. Charles Darwin in England, who appreciated Bachman's efforts to defend monogenesis, utterly rejected his defense of slavery.[8] In France, Pierre Paul Boca appreciated the book's arguments for polygenism but, with Darwin, considered Nott and Gliddon's defense of black chattel slavery reprehensible.[9] Even before Josiah Nott approached Henry Hotze, geographical location had conditioned responses to the case he and Gliddon wanted to make.

For Nott, Hotze seemed the ideal person to advance the arguments in *Types of Mankind*. For that purpose, he hoped that the Swiss immigrant could translate a book recently published by the Comte de Gobineau, a French diplomat who had titled his lengthy argument for human diversity *Essai sur L'inégalité des Races Humaines*. To Gobineau, the history of civilizations showed clearly the separate origins and continued existence of black, yellow, and white races; the threats to peace, health, and prosperity when the races mingled; and the relevance of racial history to the discontents of the modern West.[10] Hotze, who seems to have been already captivated by his personal acquaintance with Nott and his own reading of works from American ethnologists, readily agreed. Nott, we can speculate, was looking for respected foreign aid to support the conclusions of *Types of Mankind,* but support that featured general historical study rather than empirical investigations. He may also have been looking for a voice that advanced his theory of racial difference but muted his own biting criticism of traditional biblical interpreters. And so an immigrant from Switzerland set out to show how arguments from France supported theories of racial hierarchy in the United States.

The result appeared in 1856 from the same publisher, Lippincott in Philadelphia, who had brought out *Types of Mankind.* This new arrow in the American ethnological quiver contained a lengthy introduction by Hotze, his translation of the first volume of Gobineau's work (with promise of a later translation for the *Essai*'s last three volumes, which was never fulfilled), and an appendix from Nott on the question of polygenesis versus monogenesis. The title of this American translation, *The Moral and Intellectual Diversity of Races,* also reflected Nott's and Hotze's own concerns more directly than had Gobineau's original.[11]

Hotze's "editor's preface," which extended for more than one hundred pages, began with a direct address to an American public that he knew to be overwhelmingly Protestant and hence thoroughly convinced of the Scriptures' bedrock veracity. Despite what might have been assumed about Gobineau's intent from the translation's title, Hotze claimed that the work did not intend "to re-agitate the question of unity or plurality of the human species."[12] That was a question, as Hotze declared at the end of his preface's long second sentence, that "the majority of [his American] readers considered satisfactorily and forever settled by the words of Holy Writ." Instead, Hotze described the book as documenting "the leading mental and moral characteristics of the various races of men which have subsisted from the dawn of history to the present era, and to ascertain, if possible, the degree to which they are susceptible of improvement." Not theology, but now Gobineau's history, would support the science so profusely exhibited in *Types of Mankind* to further demonstrate that within the human family the various races differed "not only in degree, but in kind."[13]

Hotze's effort to reassure traditional believers that unbridgeable gaps between the races did not violate standard Christian teaching summarized what Gobineau himself wrote, at least roughly. To the Frenchman, "the only solid scientific stronghold of the believers in the unity of species" was the "prolificness of human hybrids."[14] In other words, only the fecundity of unions between Caucasians and Negroes, and among other ethnicities, spoke against what Gobineau considered overwhelming evidence for differing human species. Yet where Hotze sidelined the entire debate between either the unity or plurality of human species, Gobineau addressed it directly. As, in the words of his biographer, "a nominal Catholic," Gobineau "took pains to reconcile the fundamentally polygenetic implications of his racial theory with traditional Christian teaching on creation."[15] He thus conceded that although scientific evidence pointed to the plurality of human species, he was held back from that conclusion by "another argument which, I confess, appears to me of greater moment: Scripture is said to declare against difference of origin. If the text is clear, preemptory, and indisputable, we must submit."[16]

Despite these assertions by Hotze in his introduction and by Gobineau in the text itself, Nott in his appendix boldly claimed that "the Bible should not be regarded as a text-book of natural history." In his view, "none of the writers of the Old or New Testament give the slightest evidence of knowledge in any department of science beyond that of their profane contemporaries." As a result, he articulated a position that would later be argued at great length by John William Draper and Andrew Dickson White about an ineradicable conflict between science and theology: study of "the natural history of man is a department of science" that should be liberated from a history long "stifled by bigotry and error."[17] Nott, it seems, could not resist extending his relentless campaign against conventional religious opinion.

By contrast, Hotze, hoping to avoid theological conflict, wrote in his "analytical introduction" that "whether these races are *distinct species* or *permanent variations* only of the same, cannot affect the subject under investigation." For his purpose, it was most important to substantiate the permanent "diversities among the various branches of the human family." He clinched his case with a biblical word, though changing Jeremiah 13:23 from a question to a statement and transforming the prophet's observation about Israel's moral intractability into a proof text for racial differentiation: "The Ethiopian cannot change his skin."[18]

We are in no doubt about Hotze's intention to address a Protestant audience located in the United States because of what he wrote to Gobineau in defense of the translation. In a letter timed to arrive shortly after the latter received a copy of the translation, Hotze explained why Gobineau's "argumentative subject" had to be treated differently in the United States than in France. Since Americans, he reminded Gobineau, were "proverbially the busiest, most hurried, nation of the world," things had to be spelled out for them clearly. That cultural reality, in turn, required "certain alterations" in the original text. To Gobineau, Hotze also brought up the overarching circumstance that made such alterations necessary and specified two of the changes he had made. When Gobineau read the translation, he realized that Hotze had told him only part of the story.[19] That overarching circumstance was slavery, which Hotze informed the Comte had already led to bloodshed "in a newly formed territory" (Kansas) and had created "the rock upon which the vessel of state will wreck one day, perhaps ere long." To be sure, Hotze acknowledged that Gobineau had not written about slavery, but he knew that the Comte would certainly understand "the intimate connection of the questions you agitate and those which make this so-called Union anything but what its name implies." In Hotze's America, race meant slavery and slavery meant race.

Hotze's first "alteration" concerned the need to reassure American readers, as he had done at the outset of his preface, that diversity of races did not undermine Scripture or necessarily mean polygenesis. Hotze may have been thinking of William Archer Cocke when he explained that "the united Protestant Churches, but especially the Presbyterian, are bitterly opposed to the slightest intimation of original diversity." Cocke, in harshly attacking Nott and Gliddon's *Types of Mankind* in the *Southern Literary Messenger*, had stated his case unequivocally: "If there are distinct species of Man, then the Bible is untrue; if there are other races than the descendents [*sic*] of Adam, they are free from the penalty 'of man's first disobedience' and the tragic scene of Calvary but a mockery and a delusion."[20] Hotze, therefore, tried his best to preserve Gobineau from "the slightest suspicion of what is called infidelity" in a translation published where "we are a very religious people, and the pulpit, in some form or other, exerts a much more potent influence than it does in Europe."

Whether Hotze succeeded can be questioned, since American readers had to be confused when they read the different parts of Gobineau's work that Hotze accurately translated. On the one hand, and countering a prejudicial interpretation of Scripture deeply ingrained in American imagination, Gobineau denied that the so-called curse of Ham from Genesis 9 had anything to do with dark-skinned people: "The pretended black color of the patriarch Ham rests upon no other basis than an arbitrary interpretation." On the other hand, and with a claim that strongly supported the racial differences postulated by defenders of slavery—and that also threatened to bring down the charge of infidelity—Gobineau also wrote that "the Bible speaks of Adam as the progenitor of the white race."[21] While Gobineau claimed that he accepted biblical teaching on the unity of the human species, his interpretation of early Genesis concerning Adam strongly suggested that he belonged, as David Livingstone has cataloged him, among the advocates of "polygenetic co-adamitism."[22] For opposite reasons, the usual defenders of either monogenesis (like Cocke) or polygenesis (like Nott) considered it nonsense to assert both the unity of the species and the existence of humans outside of Adamic descent. For them, it had to be either the Bible or science, not both, as Gobineau seemed to say.

The second "alteration" that Hotze mentioned specifically was the decision he (or probably Nott) had made to cut entirely the fifteenth chapter of Gobineau's original. That chapter advanced a pessimistic theory arguing that as the races amalgamated over time, language degenerated, and this degeneration pointed toward the ultimate extinction of civilization itself.[23] Hotze told Gobineau that any such negativity about the human future "would have precluded the book from the very slightest chance of success in the United States." Americans approached the future with altogether too much hopeful expectation for such a thesis to fly. Hotze wanted Gobineau to know that he had carried out the translation by attending to American intellectual geography.

Gobineau, to say the least, was not pleased. In his own mind, he had extended the great effort required to write his book with purposes very different from those pursued by Hotze, Nott, and the American ethnologists. The *Essai* represented a lugubrious cri de coeur about the downfall of civilization, which he viewed as caused by revolution, democracy, and the decay of Europe's nobility. For him, race mixing explained nothing less than the degeneration of humanity as a whole. When a copy of Hotze's translation finally arrived, Gobineau's response mingled irritation with bemusement. To one correspondent, he wrote: "Do you wonder . . . at my friends the Americans, who believe that I am encouraging them to bludgeon their Negroes, who praise me to the skies for that, but who are unwilling to translate the part of the work which concerns them?"[24]

Authorial intent, however, meant little to Hotze and Nott, who apparently had been responsible for deleting the parts of the *Essai* not fit for American eyes. So it was that the translation also cut the long dedicatory epistle to George V, king of Hanover and member of the British royal family, in which Gobineau singled out *révolutions*, legal revisions (*renversements de lois*), and undue deference to common people (*le vulgaire*) as prime causes of civilization's decline.[25] Similarly, Gobineau's account of language exuded pessimism because of how that account supported the overall thesis of his book.

On slavery, Hotze, in fact, misspoke when he told Gobineau that his *Essai* never alluded to the subject. Actually, Gobineau had written on the situation of African-descended people in Saint-Domingue (Haiti) and Cuba, which included some notice of slavery. While the translation kept in Gobineau's dismissive comments about the barbarism of blacks on Saint-Domingue, it cut his commendation of what had happened in Cuba, where emancipated slaves had done quite well ("*les esclaves affranchis y ont mis bon ordre*").[26]

Again, although it was not a major theme in his book, Gobineau included North America in his prediction about the degeneration of civilization. Close comparison of the original and the translation would be required to identify the several places where Gobineau's dismissive comments were suppressed, but Michael Biddiss summarizes the differences clearly: Gobineau "in no way saw his work as commending either the slaveholders or those who believed in the myth of a great American future."[27] Gobineau, in other words, was explaining why, from his angle in postrevolutionary France, human civilization had entered into terminal decline. For him, North America was important since, by the 1850s, the mixing of ethnic streams had become a pronounced feature of life in the United States. For Gobineau, the mingling of Celtic, Gallic, Anglo-Saxon, and Germanic strains predicted the decline of civilization just as much as black-white amalgamation.

Nott, with Hotze, did not particularly care what Gobineau had in mind. They were publishing for an American place. It was enough that his *Essai* defended the racial differences that for them legitimated slavery. Hotze, as he told Gobineau, had needed to "consult . . . the spirit of the nation for whom I was writing." In the next phase of his career, he would maintain the same commitment to consultation but now for other audiences in different places and addressing other nations.

Hotze to London

After his translation of Gobineau's *Essai* was published, Hotze continued to move in elite American and especially Southern circles. The city of Mobile, Alabama, sent him as a delegate to an important Southern commercial convention in 1858. He served briefly as a secretary to the American diplomatic mission in Belgium, and he worked as an associate editor at the *Mobile Register*. This strongly Democratic (and hence proslavery) newspaper nonetheless opposed Southern secession because its editor, John Forsyth, remained foursquare for the Union. Little of that loyalty wore off on Hotze, however: when active conflict broke out in April 1861, he immediately enlisted in a Mobile volunteer regiment.

The assignment that eventually sent him to a new place and led to his advocating for the Confederacy in many places came in September 1861 when the Confederate secretary of war dispatched him to London for the purpose of purchasing arms and financing Southern agents in Europe. During that visit, Hotze became convinced that the Confederacy needed full-scale publicity to secure European, and especially British, support for its independence struggle. Upon returning to America, his convincing advocacy led the new Southern government to appoint Hotze as its "commercial agent" in London. When Hotze returned to England in early 1862, he first published articles in several British newspapers. Then, in April, on his own initiative and with funds from Southern supporters in England, he established a weekly newspaper that would become his main vehicle for influencing European opinion and, more specifically, securing British recognition of the Confederacy. The first issue of the *Index* appeared on May 1, 1862: the last was dated August 12, 1865, several months after the collapse of the Southern war effort. The *Index*, which gained a subscriber list of more than two thousand, seemed to make a difference in winning sympathy for the Southern perspective, especially among an elite Tory readership already inclined to view the North as an aggressively rambunctious republican threat to British interests in general and to Canada in particular.[28] The weekly's tightly printed sixteen or eighteen pages offered a Southern slant on the course of the war, provided considerable commentary on British and European affairs, reported extensively on trade and economic developments, and, as the American conflict intensified, devoted more and more columns to listing the Southern dead and wounded from military engagements. The page or page and a half of advertisements found in each issue suggested the appeal of its editorial stance to British trade, commercial, and publishing interests.[29]

For our purposes, it is noteworthy that the racialist commitments that had made Hotze so eager to translate Gobineau's work found their way into almost every aspect of the *Index*'s propaganda. In particular, as he had reported to Gobineau on how he adjusted the *Essai* for an American readership, so Hotze as the Confederacy's agent constantly kept his target audiences in view. In Britain, he addressed a readership alert to the specifically religious claims of the Confederacy. In pursuit of his aspiration to influence broader European opinion, he helped other authors address both anticlerical and

Catholic readers in Italy. For France, he wrote a pamphlet himself that appealed to that nation's imperial prospects. This targeting suggests a geographical analogy to what Quentin Skinner once described as the ideological "tailoring" of language from one conceptual sphere to another.[30] They represented the author's different efforts aimed at different readers in different nations as well as different places.

Hotze's involvement in his London locale was more complicated. His years there nicely illustrate David Livingstone's contention that "spaces must not be thought of as "givens," as mere "containers" inside which human activities take place," but as "social productions . . . constituted by social life in such a way that "mental space" and "material space" are brought together."[31] In London, not only did Hotze target what he wrote for specific audiences, but also the audiences with which he personally associated exerted a material influence on the expression of his convictions. Place shaped him, in other words, even as he directed his message to different places.

The shift over time occasioned by that two-way flow of influence has been well summarized by Robert Bonner: "Both Hotze's private dispatches and his public statements indicate a conscious move from an earlier emphasis on the white South's Christian piety and martial heroism to a consideration of how its defining system of slavery exemplified the scientific principles of racial anthropology."[32] That shift occurred amid Hotze's constant effort to reach audiences with a fixed ranking of priorities: at the most superficial level conventional religious beliefs, much deeper the Confederacy and its slave system, and most fundamental of all the principle of racial inequality.[33] Yet if these priorities were fixed, the arguments he deployed in any one piece of writing could range very widely—from an appeal to constitutional principle and ideals of manly honor; through common sense, general human experience, and history; to Christian sensibilities, the Bible, Providence, and, increasingly, science. A focus on how and to whom Hotze directed his religious and scientific arguments neglects many other aspects of his dedicated labors on behalf of the South. It has the advantage, however, of underscoring the salience of intellectual geography for his brief but energetic propaganda career.

Scottish Clergy and a London Anthropologist

That spatial salience was no better illustrated than by two major publishing events from 1863. In the June 11 issue of the *Index*, Hotze ran the entirety of an "Address to Christians throughout the World by the Clergy of the Confederate States of America."[34] Although the *Index* was regularly filled with material reprinted from other papers, extensive reports of speeches, long book reviews, and many columns devoted to verbatim publication of letters from near and far, only rarely did it offer readers the complete text of a separately published pamphlet or address. It did so again, however, in its November 26 and December 3 numbers when it printed a recently delivered lecture titled "The Negro's Place in Nature."[35] Together, the circumstances surrounding the publication of these two documents highlight the fixity of Hotze's foundational commitments, the flexibility of his arguments, and the influence of local geography on the arguments he chose to deploy.

Ninety-seven Southern ministers subscribed their names to the "Address to Christians throughout the World," which came from a conference in the Confederate capital, Richmond, Virginia. Although Presbyterians (forty-six, and of four different varieties) and Baptists (twenty-four) predominated, enough others signed from Methodist, Episcopal, Lutheran, and German Reformed churches to justify the claim that the document represented "more or less fully every accessible section of the Confederacy, and nearly every denomination of Christians"—or at least Protestant denominations. Its opening words set the tone for the whole: "CHRISTIAN BRETHREN,—In the name of our Holy Christianity, we address you . . . respecting matters of great interest to us, which we believe deeply concern the cause of our Blessed Master. . . . We speak not in the spirit of controversy, not by political inspiration, but as the servants of the Most High God we speak the 'truth in love' [Ephesians 4:15] concerning the things which make for peace [Romans 14:9] . . . above all, for the sake of our Redeemer's Kingdom." The first half of the pamphlet was devoted to justifying the war as a "defence of our liberties" in the Southern states' attempt "to secure their own rights."

In the pamphlet's second half, the clergy addressed European concerns about slavery. Their defense began with a denunciation of Abraham Lincoln's Emancipation Proclamation of January 1, 1863. The ministers asked Europeans to envision "the bloody tragedy that would appal [*sic*] humanity" if the proclamation were implemented in the Confederacy, especially since "the slaves would suffer infinitely more" than the white population from the race war the pamphlet envisioned. Although the ministers foreswore "a full discussion of this whole question of Slavery," they nonetheless devoted the rest of the pamphlet—"with our hands upon the Bible, at once the sacred chart [*sic*] of our liberties and the foundation of our faith . . . in the name of Him whose we are, and whom we serve"—to defending the system. Their defense began by quoting a passage from Luke 1:2–3 that affirmed the gospel writer's personal knowledge of what he recorded. Similarly, they claimed to testify "with all the facts of the system of slavery in its practical operations before us, 'as eyewitnesses and ministers of the Word, having had perfect understanding of all things.'" Their claim that "the relation of master and slave among us, however we may deplore abuses in this, as in other relations of mankind, is not incompatible with our holy Christianity," rested on extensive considerations of Providence, backed by briefer reference to Scripture directly.

According to the "Address," slaves arrived in America only because "Divine Providence has brought them where missionaries of the Cross may freely proclaim to them the word of salvation." Critics, disparaged earlier in the pamphlet as "philanthropists," erred in promoting "Northern fictions" about the wretchedness of servitude, since slaves in the South were "prosperous and happy." This fact, moreover, proved that "the practicable plan for benefiting the African race must be the Providential plan—the Scriptural plan." Abolition, by contrast, "we regard as an interference with the plans of Divine Providence." In a lengthy footnote, the "Address" counted half a million "coloured communicants" in the South, or the same percentage as among whites. This result proved that God had "blessed us in gathering into His Church from the children of Africa more than twice as many as are reported from all the converts in the Protestant Missions throughout the heathen world." For those who still doubted, the biblical witness was definitive. From the white South's well-filled storehouse of apposite scriptures, the ministers chose from 1 Timothy 6:1–5. This begins, "Let as many servants as are under the yoke count their own masters worthy of all honour" and includes the malediction "if any man teach otherwise, and consent not to wholesome words, even the words of our Lord Jesus Christ, . . . He is proud, knowing nothing."

Although Hotze offered no editorial comment when he published the "Address" in the *Index*, he was far from passive in promoting its distribution. In a report to the Confederate secretary of state, Judah Benjamin, on July 23, 1863, he wrote that "the address of the Southern clergy . . . had produced excellent effects" wherever it could be publicized. That publicity he provided himself not only by seeing to its separate publication but also by arranging "that a copy shall be stitched up under the same cover with this or next month's number of every respectable religious publication, as also the two leading political reviews, the Quarterly and the Edinburgh." He expected to disseminate "a quarter of a million copies of the address," which would mean that "it would be brought under the eyes of between one and two millions of readers in every part of the world where the English language is spoken."[36]

Hotze later defended the "Address" in his newspaper when a thousand Scottish ministers signed a public denunciation of the Confederate document. In that spirit of open inquiry he had announced from the start of the *Index*, Hotze published this "Reply to the Address of the Confederate Clergy" in the November 5, 1863, issue. Its authors had nothing but scorn for the Confederate document: "In the name of that holy faith and that thrice holy name which they venture to invoke on the side of a system which treats immortal and redeemed men as goods and chattels, denies them the rights of marriage and of home, consigns them to ignorance of the first rudiments of education, and exposes them to the outrages of lust and passion—we most earnestly and emphatically protest." The Scots claimed to stand with "the whole of enlightened Christendom" as they denounced the "apologists for slavery, attempting to shelter themselves and it under the authority of God's word and the Gospel of Jesus Christ."[37]

In the same issue, Hotze replied to this rejoinder with a show of rhetorical balance, which had been the stated posture of the *Index* from its inception. He first complimented the Scots on drafting a document "moderate in tone" in contrast to other Britons who had excoriated the Confederacy more harshly. Hotze did charge the Scottish ministers with "injustice to their fellow Christians in the Confederate States" but "hasten[ed] to absolve them of any willful intention of so doing." His positive apology agreed that "abuses" marred the Southern slave system, but then he claimed the Confederates would also "deem [it] a fearful sacrilege of God's Word to justify such crimes." It was not "slavery in general" that the Confederate clergy defended, or "the horrors of the slave-ship, and more recently from the novels of Mrs. Beecher Stowe, and the sensational fiction of a depraved and debauched press." It was, instead, a labor system that should more accurately be called "apprenticeship for life." According to Hotze, the Scottish denunciation deliberately misread what had been, in effect, an appeal for "the prayers of their fellow-Christians for a country in deep distress and the agonies of a great calamity." Had the Scots read carefully, they would have realized that Confederate ministers were themselves advocating for "the material and spiritual interests of the African race," and that they, too, were "very far" from "regarding [slaves] as mere chattels to whom the rights of marriage and of home might be denied, and whom they were willing to expose to the outrages of lust and passion."[38]

Despite the respect, moderation, and piety of Hotze's reply, his belief in the inequality of races shows through unmistakably. Not only did he simply equate slavery and "the African race," but he also defended the suggestion that Southern slavery be labeled "apprenticeship for life" with a categorical racial distinction: in his words, an accurate reading of the South's situation meant considering "a whole race, instead of individuals only," as subject to this lifelong "tutelage."[39] To the Scottish Christians, he spoke as a Confederate Christian so that by this means he might win over at least some to the principle of permanent racial differentiation.

In the October 29 issue of the *Index*—one week before Hotze published the reply of the Scottish ministers and his assessment of that reply—the paper featured a large advertisement for the *Anthropological Review*, a new journal published by the Anthropological Society of London. The notice promoted the journal as "a repository of facts, an arena for discussion, and a medium of communication between Anthropologists and travelers all over the world . . . irrespective of party or personal feelings" that would "promote the study of Man" while examining "the laws of his origin and progress."[40] The advertisement not only announced a new front in the propaganda war but also intimated the way that residence in London shaped Hotze's contribution to that war. David Livingstone has suggested that "*where* ideas and theories are encountered conditions *how* they are received."[41] In the case of Hotze's writing for the *Index* in 1863, it is obvious that *to whom* ideas and theories were published conditioned *how* they were presented.

Although London's Anthropological Society promoted itself simply as a contributor to scientific advance, its founding shortly after the promulgation of Lincoln's Emancipation Proclamation also hints at a political purpose.[42] The moving spirit of the society, the physician James Hunt, consolidated several lines of influence in bringing it into existence, but the one that attracted Henry Hotze played to his own most basic conviction. In a report to the Confederate secretary of state on August 27, 1863, Hotze explained that "a new scientific society . . . had been formed in London" with the express intention of "exposing the heresies that have gained currency in science and politics, of the equality of the races of men." Hotze seemed especially pleased that Hunt had offered him a seat on the new society's council. According to Hotze, Hunt told him: "You should and must take a strong interest in our objects, for in us is your only hope that the negro's place in nature will ever be scientifically ascertained, and fearlessly explained."[43] From the time of this invitation, Hotze's public advocacy underwent a noticeable shift. He did not give up appealing to religion, yet, as Hotze grew closer to the Anthropological

Society, so also did his reliance on science increase for his support of the Confederacy and his dedication to human inequality.

The *Index*'s two-part publication of Hunt's lecture "The Negro's Place in Nature" marked this shift. Coming as it did in the first year of the Anthropological Society's existence and only a few months after Hotze and two other Confederate agents had joined the society's council, it announced the ideological marriage of a British scientific upstart and the embattled advocates of the American South.[44]

Hunt's new organization arose out of a schism in the Ethnological Society of London, which itself had been created in the 1840s as an offshoot of the Aborigines' Protection Society, a body founded by Quakers and evangelical Anglicans, with Thomas Fowell Buxton, the parliamentary successor of William Wilberforce, in the lead. These reformers had mobilized out of concern for the natives of India and Africa who suffered from the triumphs of the British imperial juggernaut. The Ethnological Society sponsored careful study of humankind around the globe but did so with sturdy humanitarian intentions. Although its members included some who promoted polygenesis, often in the form of hypothesizing the existence of humans before Adam and Eve, most in the society were monogenists who explained human difference as the result of climate, education, or what later would be called culture.[45] Charles Darwin, a prominent member of the Ethnological Society, was distinguished from most of his fellow members by explaining monogenesis as the result of natural evolutionary processes instead of direct divine creation.

Against these humanitarians, a minority in the society, led by James Hunt, pushed back. They advanced under the banner of Thomas Carlyle, who in a much publicized essay on "the Negro question" from 1849 had argued that disparities of wealth and civilized advance were best explained by permanent racial differences and not by the workings of Adam Smith's morally neutral marketplace. Dissenters in the Ethnological Society also considered the growth of the British Empire as a nearly unalloyed good that benefited rather than harmed people of darker skin. In addition, such people saw themselves as the intellectual wave of the future because in studying the various peoples of the world they appealed to scientific fact rather than humanitarian impulses or to biblical teachings.

With this background, the American Civil War, as well as the earlier stimulus given to polygenesis by Gliddon and Nott's *Types of Mankind*, created the stuff of a natural alliance. Hunt found a new venue for his work in the *Index* and other outlets under Hotze's influence. For his part, Hotze, as described by Adrian Desmond and James Moore, "eagerly tied" the new society's "science of Negro inferiority to the Confederate cause."[46] It was the grafting of American interests onto a British organization that led to the appearance of Hunt's lengthy two-part article in the *Index* in late November and early December 1863.

As he described "the Negro's place in nature," Hunt claimed the scientific high ground: he spoke "not to support some foreign conclusion, but to endeavour to ascertain what is the truth by careful and conscientious examination and discussion of the facts before us."[47] Against the absurd notion ("I am astonished") that "all races have the same intellectual, moral, and religious natures," Hunt declared "the real fact." Rather than a common humanity acted upon by uniform influences, "the greatest differences" divided "each race" from the others. In the particular case under consideration, "the assertion that the Negro only requires an opportunity for becoming civilized, is disproved by history" and also by anatomy, physiology, and psychology.[48] After giving an unprecedented twelve-column exposure to Hunt, Hotze took three more columns in the next issue of the *Index* (December 10, 1863) to cushion religiously, but also to underscore, the lecture's conclusions. Cushioning involved Hotze's description of humankind (and human differences) as "the noblest work of the Creator." It involved his assertion that blacks possessed many "good qualities," especially the capacity for "receiving and profiting by the Christian religion." And it crept into his defense of the South when he observed "the wonderful . . . progress of the coloured race under the mild, humane, and Christian rule of the Southerners."[49]

Yet, underlying this religious superstructure was an adamantine deep structure of scientific racism. Hunt had offered a "truly philosophic paper" that could not "fail to command the critical attention of the scientific world." His "earnest, single desire to discover the truth" demonstrated that the same "marvelous progress of physical science since the days of Bacon" would now be replicated "with respect to moral and social science." Grounded firmly on that scientific basis, the conclusions proceeded categorically:

- "the brain of the negro is of smaller capacity";
- "the negro has no history";
- "the negro is indubitably inferior in intellect";
- and with Africa in view, "it is utterly impossible to exaggerate the savage barbarity and the utter degradation of the negro at home."

Hotze, the propagandist, then put his science to work by asserting that "the best condition of life in which the negro has until now been placed is that in which he is found in the Confederate States." But also, as an intimation for what his stance would become once the military tide turned, he conceded that "perpetual slavery" might not be the only destiny for "the negro," but "surely . . . the guidance and the intellect and the will of the white man are indispensable to him."[50]

In Mobile, Hotze employed ethnographic science to support American slavery, but then in London he used it to make propaganda for the slaveholding Confederacy. With the passage of time, science for Hotze became more important for defending racial hierarchy than for defending either slavery or the South. Science, per se, never existed outside of time and beyond space.

Religion, Science, and Racial Difference

Throughout the entire career of the *Index*, Hotze appealed to religion and science—as well as to history, common sense, and the American Constitution—to defend ineradicable racial difference. Yet as the rapid survey that follows indicates, the proportions within his defense did shift after he allied Confederate interests with the scientific ambitions of Hunt's Anthropological Society. As his responses to the Scottish clergy had shown, Hotze kept ever in view the national locations to which he directed his work. After his London location enabled him to league with the Anthropological Society, his own locale shaped him as he hoped his propaganda for other places would shape opinions in those nations.

Throughout its first two years, the *Index* regularly featured conventional religious pieties in its defense of the Confederacy and its slave system. In its second month, Hotze published a letter "from an Englishman in the South" who proclaimed "the magnitude of nature's great law . . . of subjection" with a heavy dose of Providence: "the ways of God . . . 'His ways are not our ways' [quoting Isaiah 55:8] . . . the Great Jehovah."[51] In November, Hotze published an extensive leader devoted to a sermon preached two years earlier by Benjamin Palmer of New Orleans that had explained slavery in the South as "a solemn Providential trust." The slave system, as well as the resulting civil conflict, as Hotze summarized this Presbyterian sermon, was not of the South's own creation but reflected "the fiat of the Almighty."[52] Two months later, the *Index* reprinted a much-debated column from the London *Times* that criticized Northern abolitionist ministers such as George Cheever and Henry Ward Beecher by noting that "although they preach with the [B]ible in their hands, . . . in that book there is not a single text that can be presented to prove slavery unlawful."[53] The *Index* offered its own opinion a month later by describing Lincoln's Emancipation Proclamation as "condemned by the laws of God and man," as well as by the United States' Constitution. The same editorial asserted that due to the American slave system, "in half a century has the fierce sanguinary African become a Christian labourer."[54]

So the religious notes continued, although with considerably reduced frequency, through 1864 with occasional references to "the Divine Word" or "the facts as God had made them."[55] After the collapse of the Southern war effort and the beginning of Reconstruction, Hotze again took note of religion but only to bewail abuses. In the *Index*'s penultimate issue, he heaped scorn on efforts to transform slaves into citizens. He was particularly appalled by reports concerning "philanthropic . . . apostles of freedom and equality, who in the presence of Federal officers preach to the negroes in churches and schools that they are the superior race, that God created man black, but that Cain, his face blanched by fear and his hair straightened by his rapid flight from the wrath of the Almighty, is the father of the white man."[56] Although references to Providence continued in Hotze's correspondence, public appeals to religion faded almost completely away in the last years of the *Index*.[57]

When Hotze added science to his arguments in the *Index*, he adopted the same strategy for Britain that he had earlier used in presenting Gobineau's work to the American public. Discoveries from science, he again wrote, should be regarded by the faithful not as replacing trust in revelation but as explaining the work of God with greater precision. In lengthy reviews of his (and Nott's) Gobineau translation and, later, of an early essay by James Hunt, Hotze did give more credit to divine agency than either Josiah Nott or James Hunt had done in the works he reviewed. In both reviews, however, Hotze left no doubt about what he most wanted to say. He told his readers that Nott's appendix to the translation of Gobineau's *Essai* contained "the result of the most recent discoveries in his own and cognate branches of science," which demonstrated irrefutably "the ineradicable character of hereditary qualities." Yet, according to Hotze, religious believers should not worry, since "Revelation assures us that the sins of the fathers are visited on the children; and science, as usual, confirms and illustrates abundantly the brief assertion of Revelation."[58]

Similarly, in his review of Hunt's contribution to the first issue of the Anthropological Society's journal, Hotze led off with his essential point: "The most dangerous idea of modern times and that which, unconsciously to the majority of those who accept it, underlies nearly every social, political, and religious heresy which mars our civilization, is the dogma of the equality of man." Again, however, believers had no cause for concern. As he had for Gobineau, Hotze insisted that Hunt's scientific conclusions remained agnostic about the question of polygenism or monogenism since the science presented only an "apparent contradiction to Holy Writ." In reality, "whether the Great Architect primarily ordained the existing differences between distinct varieties of man, or whether He subsequently effected the same object by such means as He saw proper, science has to deal only with the facts as she finds them."[59] Science had become Hotze's main weapon to slay the monogenists, but unlike Nott and Hunt, he did not want to abandon entirely the religious armament that had served him well in his intellectual campaign for British popular opinion.[60]

Location, Location, Location

Even as Hotze shifted his rhetoric away from religious language toward an evocation of science, it was obvious that racialism grounded his entire effort. As illustrated by the *Index*'s reviews of his own and Hunt's publications, Hotze consistently followed long-standing lawyerly practice: to those with a concern for law (revelation), argue the law; for those attuned to the facts of a case (science), argue the facts. The verdict, not the arguments, remained the goal. In all of Hotze's labors, however, intellectual geography played a continuing role. Spotlighting two more features of his work illuminates the further significance of place. One was the support that national essentialism lent to his ethnic essentialism; the second was his assistance to two authors who contended for racial differentiation before two different continental audiences.

Throughout his work for the Confederacy, Hotze regularly spoke of nations as possessing the same kind of differentiated character as he perceived among the races. Before Hotze's *Index* review of his own Gobineau translation arrived at its conclusion—that "Negro and European are as much distinct species as are horse and ass, or hare and rabbit"—it took a lengthy detour past national characteristics. This detour expatiated at length on how "the Frenchman and the Irishman . . . have nothing in common, except an ancestry which was divided before the memory of man." In the same review, he also specified the characteristics that differentiated the English, the French, "Celtic nations," "the Turk," "the Greek," "the Neapolitans," "the Piedmontese," and "the Lombards" from one another.[61] A similar recourse to national stereotypes appeared regularly in his correspondence with Confederate Secretary of State Judah Benjamin. The English, he wrote in August 1862, displayed an "intense selfishness" that "overshadows all other national characteristics."[62] In a long report from mid-1863, he spelled out the national characteristics that shaped the differences in antislavery sentiment found in England, France, Germany, and the continent more generally.[63]

The next year, the same reasoning informed a pamphlet he wrote to encourage France's imperial outreach into Mexico, a move that would have aided the Confederacy's goal of staving off Union victory. This pamphlet depended wholly on national essentialism. British national character, it began, was due to "l'accident géographique" that gave this people "une situation insulaire." France, by contrast, enjoyed a distinction because of its national mental superiority. "Aucun pays de l'Europe n'a produit les fruits de l'intelligence en aussi grand abondance, n'a regné aussi longtemps en sourverain absolu dans la domaine de la pensée." As detailed by Hotze, the French were "serious, sober, meditative, not very fanciful, even less eccentric, and eminently creative" (sérieux, sobre, meditatif, peu fantasque, encore moins excentrique, et éminemment créateur). Above all, "L'organisation, qui est la plus haute expression de l'ordre, est instinctive chez le Français."[64] On the basis of these national characteristics, Hotze then explained why colonizing Mexico would give the distinctive French genius a much-deserved outlet.

In a note to Secretary of State Benjamin, Hotze acknowledged modestly that "with the assistance of an able translator I was the secret author" of the pamphlet.[65] With no modesty at all, he then reviewed his own pamphlet in the *Index* as "a very remarkable work of mysterious authorship . . . replete with vigorous thought and so suggestive." This review summarized the pamphlet's argument for French support of Emperor Maximilian's colonial escapade in Mexico, ascribed the superfluity of European intellectual energy to "the wise plan of Providence," and cited the early history of Canada to demonstrate that "as colonizer the Frenchman possesses a superiority over the Englishman in certain very essential attributes of character."[66] Hotze did not link his depiction of essential national characteristics directly to his racial essentialism. Yet in the same way that he marshaled different arguments (religious, scientific, historical, constitutional) for various demonstrations of white superiority and black inferiority, so he employed a discourse of national essentialism in different ways depending on whom he was addressing. Place again shaped what he wanted to say and where.

Audience also became a factor in the assistance Hotze provided to two very different Italian publications designed to win support for the Confederacy cause on the continent. As part of his wide-ranging duties as the South's "commercial agent," Hotze in May 1864 facilitated the journey to Rome of Catholic Bishop Patrick Lynch of Charleston, South Carolina, who had been enlisted by Jefferson Davis to promote the Confederate cause in Europe.[67] The major product of the bishop's trip was a defense of slavery that he published first in Italian (August 1864) and then in French and German versions as well. Although Hotze did not contribute directly to what Lynch wrote, the bishop's anonymous *Lettera di un Missionario sulla Schiavitù Domestica degli Stati Confedrati de America* deployed many of the arguments that had regularly filled the *Index*—British responsibility for introducing slavery into America, the zeal of the South for liberty, the North's unconstitutional brutality, the exaggerated nature of stories recounting slave abuse, and especially the South's humane treatment of the slaves, including the successful Christianization of many. The pamphlet most clearly reflected Hotze's views when it rang the changes on the racial determination so clearly visible in the South: "The negroes are, as a race, very prone to excesses, and, unless restrained, plunge madly into the lowest depths of licentiousness. . . . the character of the negro race . . . conforms with difficulty to the strict requirements of Christian morality. . . . [With emancipation] no race would be more relentlessly pursued than the unfortunate African, marked by his colour and features, and so different in traits of mind and habits of body."[68]

The second Italian publication that Hotze midwifed into existence was very different in tone and content. The same report to Secretary of State Benjamin in which Hotze had enthusiastically recounted his invitation to join the council of James Hunt's new Anthropological Society also included the news that he was employing an Italian in Turin on behalf of the Confederacy.[69] The result of that subvention appeared in 1864 as a substantial book from Filippo Manetta. The author, a follower of republican revolutionary Giuseppe Mazzini, had once lived in the United States; he now eagerly accepted Hotze's invitation to make propaganda for the South.[70] The preface announced that the book had only one purpose (*uno scopo solo*), which was to canvass the opinions of distinguished anthropologists from the Old World and the New, as well as past and present travelers, for what they reported concerning "al carattere morale, intellettuale e fisico della RAZZA NEGRA."[71] The work proper began with a lengthy attack on Lincoln's Emancipation Proclamation, followed by reports from Italian travelers on African physiology and psychology, quotations from African travel journals by English adventurers, and the "opinioni di celebri viaggiatori [travelers] e scienziata sulle condizoni dei Negri" in Africa, Brazil, Cuba, Haiti, and the American North and South—and all intermixed with regular pro-Confederate, anti-Union obiter dicta.[72] Manetta's conclusion came as no surprise: "Now, after philosophers, travelers, and champions of all religions and all sciences have written so much about the physical, moral, and intellectual character of the Negro race, can we hesitate to admit that it is inferior to our own?"[73] Unlike Bishop Lynch's humanitarian, political, and religious racialism, Manetta's more secular account tried to convince a different Italian readership of the scientific justification for the South's racial hierarchy. Even if Hotze's racialism informed both works, their different targeting replicated the differentiated strategy that marked his own British propaganda.

Conclusion

When Confederate defeat became inevitable, Hotze held out hope that racist hierarchy could still be preserved. To a New York correspondent in late April 1865, he wrote words probably targeted mostly to himself. If the South simply gave in to Northern occupation, the United States would stand "on the eve of a civil revolution more momentous than the war itself." The ultimate disaster would be "the Africanization of the [restored] Union." Once Northern despotism decreed "negro suffrage," national degradation would be the only possible outcome. Still, he went on, "if there is manhood and common sense enough left in the victorious section to construct a white man's government out of the smouldering ruins that negrophilism and all the other accursed isms of your section have left, I should like to have my part in the work."[74] Whatever Hotze came to think about the course of Reconstruction and the unbearable prospect of "Africanization," he never did return to the United States. Shortly after closing down the *Index*, Hotze moved to Paris, where he married the daughter of a former Confederate official and earned his living as a journalist. He died in his native Switzerland at Zug in 1887.

Hotze's general historical significance was his contribution to the scientific racism that flourished in the West during the second half of the nineteenth century.[75] With his promotion of Gobineau, his sponsorship of Manetta, and his own many articles in the *Index*, Hotze did his part to enlist the human sciences in support of racial hierarchy. At the same time, it would be difficult to imagine a career that so clearly demonstrated the effects of geography on questions linking science, religion, and society—and in so doing reinforced the key insights of David Livingstone's scholarship. At the end of his book describing the different receptions accorded Darwin's theory of evolution by religious communities in Edinburgh, Belfast, Toronto, Columbia (South Carolina), and Princeton, Livingstone concluded that "as [Darwin's] theory diffused, it diverged."[76]

In Henry Hotze's career as translator, editor, and propagandist, a different kind of divergence can be glimpsed, though one just as sensitive to place. For his efforts at defending what he considered the racialist truth about the modern world in general and the Civil War in particular, *he* was the one who used religion and science to advance his convictions with "different styles" and in different "speech spaces."[77] In this case, the author himself, and not others, created the divergent expressions that diffused his theory.[78]

Notes

1. Hotze to Gobineau, January 1, 1856, in *Henry Hotze: Confederate Propagandist: Selected Writings on Revolution, Recognition, and Race*, ed. Lonnie A. Burnett (Tuscaloosa: University of Alabama Press, 2008), 184; original in Ludwig Schemann, *Gobineaus Rassenwerk: Aktenstücke und Betrachtungen zur Geschichte und Kritik des Essai sur L'inégalité des Races Humaines* (Stuttgart: F. Frommanns, 1910).

2. Henry Hotze to Confederate Secretary of State Judah Benjamin, September 26, 1863, "C.S.A. Commercial Agency—London" (letter book), Henry Hotze Papers, 1861–1865 (MMC-0677), Library of Congress. The Library of Congress holds this letter book along with a second, titled "Letter Book of Henry Hotze, May 28, 1864-June 16, 1865." They are identified hereafter as First Letter Book and Second Letter Book, Hotze Papers. Some of the Hotze manuscripts in the Library of Congress were published in *The Official Records of the Union and Confederate Navies in the War of Rebellion*, ed. Harry Kidder White, series II (Washington: Government Printing Office, 1922), 3:505–8, 691–94, 849–51, 866–68, 915–18, 944–48, 1027–28.

3. David N. Livingstone, "Debating Darwin at the Cape," *Journal of Historical Geography* 52 (2016): 14. On the scientific, racial, and religious questions that engaged Hotze, see David N. Livingstone, "Evolution as Metaphor and Myth," *Christian Scholar's Review* 12 (1983): 111–25; David N. Livingstone, *Darwin's Forgotten Defenders: The Encounter between Evangelical Theology and Evolutionary Thought* (Grand Rapids: Eerdmans, 1987); David N. Livingstone, "Evolution and Eschatology," *Themelios* 22 (October 1996): 26–36; and David N. Livingstone, *Adam's Ancestors: Race, Religion and the Politics of Human Origins* (Baltimore: Johns Hopkins University Press, 2008). His studies that have made my own historical writing much more alert to the realities of intellectual geography also include *The Geographical Tradition* (Oxford: Blackwell, 1992); "Darwinism and Calvinism: The Belfast-Princeton Connection," *Isis* 83 (1993): 408–28; "Science and Religion: Toward a New Cartography," *Christian Scholar's Review* 26 (1997): 270–92; *Science, Space and Hermeneutics: Hettner-Lecture 2001* (Heidelberg: University of Heidelberg Department of Geography, 2002); *Putting Science in Its Place: Geographies of Scientific Knowledge* (Chicago: University of Chicago Press, 2003); "Science, Site and Speech: Scientific Knowledge and the Spaces of Rhetoric," *History of the Human Sciences* 20 (2007): 71–98; and *Dealing with Darwin: Place, Politics, and*

Rhetoric in Religious Engagements with Evolution (Baltimore: Johns Hopkins University Press, 2014).

4. Hotze, whose name also appears as Hotz, has recently been well served by an excellent article: Robert E. Bonner, "Slavery, Confederate Diplomacy, and the Racialist Mission of Henry Hotze," *Civil War History* 51 (2005): 288–316; and Lonnie Burnett's well-edited selection from his works, *Henry Hotze*. Unless otherwise indicated, bibliographical information on Hotze comes from these two sources or from the useful sketch by D. P. Crook in the *American National Biography* (New York: Oxford University Press, 2000), 11:245–47. Bonner's article is especially useful for identifying Hotze's most important publications.

5. On this work and the controversies it generated, I have been guided by Livingstone, *Adam's Ancestors*, 174–86; William Stanton, *The Leopard's Spots: Scientific Attitudes toward Race in America, 1815–59* (Chicago: University of Chicago Press, 1960), 161–74; Reginald Horsman, *Josiah Nott of Mobile: Southerner, Physician, and Racial Theorist* (Baton Rouge: Louisiana State University Press, 1987), 171–205; Adrian Desmond and James Moore, *Darwin's Sacred Cause: How a Hatred of Slavery Shaped Darwin's Views on Human Evolution* (Boston: Houghton Mifflin Harcourt, 2009), 262–66; and for more general background, see Michael O'Brien, *Conjectures of Order: Intellectual Life and the American South, 1810–1860*, 2 vols. (Chapel Hill: University of North Carolina Press, 2004), 1:215–52 (notably chapter 5, "Types of Mankind"), and G. Blair Nelson, "'Men before Adam!': American Debates over the Unity and Antiquity of Humanity," in *When Science and Christianity Meet*, ed. David C. Lindberg and Ronald L. Numbers (Chicago: University of Chicago Press, 2003), 161–82, 304–7.

6. J. C. Nott and Geo. R. Gliddon, *Types of Mankind* (Philadelphia: Lippincott, 1854), 56, 467, 260; emphasis original.

7. See Lester D. Stephens, *Science, Race, and Religion in the American South: John Bachman and the Christian Circle of Naturalists, 1815–1895* (Chapel Hill: University of North Carolina Press, 2000), 195–211.

8. Desmond and Moore, *Darwin's Sacred Cause*, 263–64.

9. O'Brien, *Conjectures of Order*, 1:248.

10. A translation of Gobineau's first tome that is more complete than Hotze's was provided by Adrian Collins, *The Inequality of Human Races by Arthur de Gobineau*, intro. Oscar Levy (New York: Howard Fertig, 1967 [orig. 1915]). Michael D. Biddis offers helpful context in *Father of Racist Ideology: The Social and Political Thought of Count Gobineau* (New York: Weybright and Telley, 1970); and Biddis, ed., *Gobineau: Selected Political Writings* (New York: Harper and Row, 1970).

11. The full title was *The Moral and Intellectual Diversity of Races, with Particular Reference to Their Respective Influence in the Civil and Political History of Mankind. From the French of Count A. de Gobineau: With an Analytical Introduction and Copious Historical Notes by H. Hotz. To Which is Added an Appendix Containing a Summary of the Latest Scientific Facts Bearing Upon the Question of Unity or Plurality of Species by J. C. Nott, M.D. of Mobile* (Philadelphia: Lippincott, 1856).

12. In a second letter to Gobineau, dated July 11, 1856, Hotze reported that "I am neither Unitarian [monogenist] or Polygenist." Although he did not care to dispute such points, he remained absolutely delighted that in his estimation Gobineau had conclu-

sively documented the "original . . . diversity of human racial types": See Burnett, *Henry Hotze*, 187.

13. Hotze, introduction, *Moral and Intellectual Diversity of Races*, vii–viii.

14. Gobineau, *Moral and Intellectual Diversity of Races*, 337.

15. Biddis, *Father of Racist Ideology*, 118, 107.

16. Gobineau, *Moral and Intellectual Diversity of Races*, 337.

17. Nott, Appendix C, "Biblical Connections on the Question of Unity or Plurality of Species," in *Moral and Intellectual Diversity of Races*, 506. See John William Draper, *History of the Conflict between Religion and Science* (New York: Appleton, 1875); A. D. White, *A History of the Warfare of Science and Theology in Christendom* (New York: Appleton, 1896).

18. Hotze, "Introduction," *Moral and Intellectual Diversity of Races*, 15–16; emphasis original.

19. All quotations directed by Hotze to Gobineau in this paragraph and the remainder of this section are from Hotze to Gobineau, January 1, 1856, in Burnett, *Henry Hotze*, 183–86.

20. Quoted in Horsman, *Josiah Nott*, 198.

21. Gobineau, *Moral and Intellectual Diversity of Races*, 338.

22. Livingstone, *Adam's Ancestors*, 159.

23. See Gobineau, *Essai sur L'inégalité des Races Humaines, Tome Premier* (Paris: Dido Fréres, 1853), 321: "Aussitôt qu'a lieu le mélange des peuples, les langues respectives subissent une révolution, tantôt lente, tantôt subite, toujours inévitable. Elles s'altèrent, et, au bout de peu de temps, meurent." Also, and on 323: "On peut poser en thèse générale qu'aucun idiome ne demeure pur après un contact intime avec un idiome différent."

24. Gobineau quoted in Biddiss, *Father of Racist Ideology*, 147.

25. Gobineau, *Essai*, ii-iii.

26. Ibid., 82n1.

27. Biddiss, *Father of Racist Ideology*, 147.

28. On the effectiveness of the *Index* as a propaganda tool, particularly in association with Hotze's work for the Southern Independence Association, see Thomas E. Sebrell II, *Persuading John Bull: Union and Confederate Propaganda in Britain, 1860–1865* (Lanham, MD: Lexington, 2014), 2–8, 57–71, 137–39, 191–200.

29. I would like to offer special thanks to Dr. Rachel Bohlmann of Notre Dame's Hesburgh Library for securing a copy of the *Index* for the library. Authors' names did not appear with articles in the *Index*, but a letter from 1864 Hotze indicated that the newspaper's many treatments of race and related questions "are always my own"; Hotze to Felix Aucaigne, January 29, 1864, *Official Records of Union and Confederate Navies*, series II, 3:1027. See Bonner, "Slavery, Confederate Diplomacy, and Hotze," 297n20.

30. Quentin Skinner, *The Foundations of Modern Political Thought, Vol. 1: The Renaissance* (New York: Cambridge University Press, 1978), xi.

31. Livingstone, "Science, Site and Speech," 72.

32. Bonner, "Slavery, Confederate Diplomacy, and Hotze," 290.

33. On the possibility that Hotze had been educated in Catholic schools and

retained some commitment to that early training, see Bonner, "Slavery, Confederate Diplomacy, and Hotze," 293n10.

34. This was the full title of the pamphlet that Hotze himself arranged to publish in London with Strangeways and Walden; the title in the *Index* (June 11, 1863, 108–10) read only "Address to Christians throughout the World." The separately published pamphlet and the newspaper's rendering were otherwise identical. The quotations that follow are from the *Index*.

35. "Dr. Hunt on the Negro's Place in Nature," *Index*, November 26, 1863, 486–87; and December 3, 1863, 501–3.

36. Hotze to Benjamin, July 23, 1863, in *Official Records of the Union and Confederate Navies*, series II, 3:850.

37. "Reply to the Address of the Confederate Clergy," *Index*, November 5, 1863, 439. As with the original address, this reply was widely distributed, as for example, in the *London Daily News*, November 3, 1863, 2, and *Leeds Mercury*, November 4, 1863, 4.

38. "The Reply of the Scottish Clergy," *Index*, November 5, 1863, 441–42.

39. "The Reply of the Scottish Clergy," 442.

40. "Now ready, the first two numbers of a new quarterly scientific journal," *Index*, October 29, 1863, 431.

41. Livingstone, "Science, Site and Speech," 73; emphasis original.

42. My account of the Anthropological and Ethnological Societies depends heavily on Livingstone, *Adam's Ancestors*, 112–14, 170–73; and Desmond and Moore, *Darwin's Sacred Cause*, 332–69 (with 331–33 especially illuminating on Hotze and the *Index*). For an account of how race, slavery, and emancipation featured in British propaganda from both North and South, see Duncan Andrew Campbell, *English Pubic Opinion and the American Civil War* (Rochester, NY: Boydell, 2003), 124–33.

43. Hotze to Judah Benjamin, August 27, 1863, First Letter Book, Hotze Papers.

44. On the addition of George McHenry and Albert Taylor Bledsoe to the council, and the election of Josiah Nott as an honorary member, see Desmond and Moore, *Darwin's Sacred Cause*, 413–14n37, and Bonner, "Slavery, Confederate Diplomacy, and Hotze," 301.

45. On John Crawford, who promoted polygenesis as a result of divine creation, see Livingstone, *Adam's Ancestors*, 113.

46. Desmond and Moore, *Darwin's Sacred Cause*, 333.

47. "Dr. Hunt on the Negro's Place in Nature," *Index*, November 26, 1863, 487.

48. "Dr. Hunt on the Negro's Place in Nature (Concluded)," *Index*, December 3, 1863, 501.

49. "The Negro's Place in Nature," *Index*, December 10, 1863, 522–23.

50. "The Negro's Place in Nature," 523.

51. "From an Englishman in the South," *Index*, June 26, 1862, 131.

52. "The South and Slavery," *Index*, November 6, 1862, 25. Palmer's effort was originally published in a volume containing sermons from both North and South preached shortly before fighting broke out: *Fast Day Sermons: Or, the Pulpit on the State of the Country* (1861). On these sermons, including Palmer's, see Mark A. Noll, "The Peril and Promise of Scripture in Christian Political Witness," in *Christian Political Witness*, ed. George Kalantzis and Gregory W. Lee (Downers Grove, IL: InterVarsity Press, 2014), 35–55.

53. "The Fanaticism of the Abolitionists (From the *Times*, January 6)," *Index*, January 8, 1863, 173. For fuller treatment of the *Times*' statement, see Mark A. Noll, *The Civil War as a Theological Crisis* (Chapel Hill: University of North Carolina Press, 2006), 106–7, 116.

54. "A Word for the Negro," *Index*, February 12, 1863, 249–50.

55. "The 'Edinburgh Review' on the Negro," *Index*, January 21, 1864, 41 ("Divine Word"); "The 'Foul Blot.'" *Index*, February 18, 1864, 165 ("God").

56. "The Negro Race in America," *Index*, August 12, 1865, 489.

57. Hotze to John George Witt, August 11, 1864, Second Letter Book, Hotze Papers: he wanted the *Index* to promote, among many other virtues, "the wisdom of Providence rather than human ingenuity." Hotze to B. Wood, April 21, 1865, Second Letter Book, Hotze Papers: He could not believe "it is the Divine will" for the defeated South to sink to "the miserable spectacle of the South and Central American Republics."

58. "The Distinction of Race," *Index*, October 23, 1862, 414.

59. "The Natural History of Man," *Index*, July 23, 1863, 204.

60. Hotze's commentary on Charles Darwin in his review of the Gobineau translation depicted Darwin's research as superficial by comparison to Nott's and claimed that Darwin's clinging to "common descent" resulted in "shipwreck" for his theories: "The Distinctions of Race," *Index*, 413 (comparison to Nott), 414 (shipwreck); Desmond and Moore, *Darwin's Sacred Cause*, 333.

61. "The Distinctions of Race," October 23, 1863, 414.

62. Hotze to Benjamin, August 11, 1862, First Letter Book, Hotze Papers.

63. Hotze to Benjamin, September 26, 1863, First Letter Book, Hotze Papers.

64. Anonymous, *La Question Mexicaine et la Colonisation Francais* (London: H. Bailliere, [1864]), 2, 10, 11.

65. Hotze to Benjamin, February 10, 1864, First Letter Book, Hotze Papers.

66. "France and Mexico," *Index*, January 21, 1864, 41–42.

67. On Lynch's journey and the proslavery pamphlet he wrote, see David C. R. Heisser and Stephen J. White Sr., *Patrick N. Lynch, 1817–1882: Third Catholic Bishop of Charleston* (Columbia: University of South Carolina Press, 2015), 94–125 and 104–6, specifically on Hotze's assistance. Hotze reported on Lynch's mission to Benjamin, June 10, 1864, First Letter Book, Hotze Papers.

68. Quotations are from a recent edition of Lynch's tract taken from his English-language drafts, which has never before been published. David C. R. Heisser, ed., "'A Few Words on the Domestic Slavery in the Confederate States of America,' by Bishop Patrick Nelson Lynch, Part Two," *Avery Review* 3 (2000): 93–123 (quotations from 97, 101, 108, respectively). Part I of Lynch's tract appears in *Avery Review* 2 (1999): 64–103. The quotations correspond to *Lettera di un Missionario sulla Schiavitù Domestica degli Stati Confederati di America* (Rome: Giovanni Cesaretti, 1864), 39, 47, 60.

69. Hotze to Benjamin, August 27, 1863, First Letter Book, Hotze Papers.

70. On Manetta, I could find virtually nothing in English. But see the informative article "Filippo Manetta un italiano sostenitore dei confederati e della supremazia bianca," *Eretica Mente*, http://www.ereticamente.net/2013/03/filippo-manetta-un-italiano-sostenitore-dei-confederati-e-della-supremazia-bianca.html (accessed October 27, 2016).

71. Filippo Manetta, *La Razza Negra nel suo Stato Selvaggio in Africa e nella sua*

Duplice Condizione di Emancipata e di Schiavi in America (Torino: Tipografia del Commercio, 1864).

72. Among these authorities was John Henning Speke, whose works on Africa would later play a significant role in the reification of Hutu-Tutsi difference in Rwanda; Manetta, *Razza Negra*, 65–66. On his place in that evil genealogy, see Timothy Longman, *Christianity and Genocide in Rwanda* (New York: Cambridge University Press, 2010).

73. Manetta, *Razza Negra*, 165: "Ora, dopo quanto hanno scritto i filosofi, i viaggiatori, e gli apostoli tutti della religione e della scienza intorno al carattere fisico, morale ed intellettuale della razza Negra, potremo noi esitare ad ammettere che essa non sia inferiore alla nostra?"

74. Hotze to B. Wood, April 21, 1865, Second Letter Book, Hotze Papers.

75. See especially Bonner, "Slavery, Confederate Diplomacy, and Hotze," 314–16; and for the general picture, see Colin Kidd, *The Forging of Races: Race and Scripture in the Protestant Atlantic World, 1600–2000* (New York: Cambridge University Press, 2006), especially 121–67 ("Monogenesis, Slavery and the Nineteenth-Century Crisis of Faith").

76. Livingstone, *Dealing with Darwin*, 197.

77. For those phrases, see Livingstone, *Dealing with Darwin*, 172, 173.

78. My thanks to Jim Moore, Robert Bonner, and the editors for helpful comments on an earlier draft, but mostly to David Livingstone for alerting so many grateful readers to the importance of place.

CHAPTER FOUR

"Made in America"

The Politics of Place in Debates over Science and Religion

RONALD L. NUMBERS

For decades historians of science—and no one more than David N. Livingstone—have been stressing the importance of place in the history of science, including its relationships with religion.[1] Other scholars have emphasized the peculiarities of science-religion interactions in the United States. Inspired by these two emphases, this chapter traces assertions of American distinctiveness with respect to science and religion from the 1840s, when such claims first became common, to the present. In exploring this topic, it is important to keep in mind that the United States is a large and diverse nation, spanning an entire continent. Reactions to scientific developments frequently varied from region to region, denomination to denomination, race to race. Here I refrain from examining this diversity, focusing instead on what observers, especially foreigners, alleged.

Before the 1840s, Europeans rarely took note of scientific activities in the United States, largely because—with the notable exception of Benjamin Franklin—few Americans produced world-class science. Much of the exploratory work in natural history had been done by foreign visitors. Some Americans, embarrassed by being the mere recipients of European knowledge, blamed their scientific inactivity on the availability of foreign literature and on feelings of national inferiority. Harvard's Oliver Wendell Holmes, for example, discerned a "fatal influence" to the growth of indigenous science emanating from the indolence created by the "fairest fruits of British genius and research [being] shaken into the lap of the American student." In 1820, an essayist for the *Edinburgh Review* asked contemptuously, "What does the world yet owe to American physicians or surgeons?"[2] He could well have added American men of science. In an effort to spur the medical and scientific communities into action, the *Philadelphia Journal of the Medical and*

Physical Sciences for several years displayed the insulting Edinburgh query prominently on its title page.

By the mid-1840s, the situation was rapidly changing. As Benjamin Silliman, the founding editor of the *American Journal of Science*, later observed: "The year 1845 mark[ed] the beginning of a new era in the scientific life of America." Among the many accomplishments he enumerated were the arrival the next year of the Swiss naturalist Louis Agassiz, a rising star in the world of international science; the opening of the Smithsonian Institution in 1846 under the direction of Joseph Henry; the revival of the US Coast Survey under the dynamic leadership of Alexander Dallas Bache, who turned it into the principal scientific institution in America; the opening by Ormsby M. Mitchel of the Cincinnati Observatory, the first public observatory in the New World; the founding of the American Association for the Advancement of Science in 1848; and the endowment in 1847 of scientific schools at Yale and Harvard. By 1860, the *New York Times* was posting on its front page: "American Science: Is There Such a Thing?"[3]

Because the first eighty years of the American Republic coincided with, first, a major religious awakening and, second, increasing conflict, much early national history has focused on religion and race. The first half of the nineteenth century also witnessed major developments in one scientific discipline after another: from geology and chemistry to physics and pathology. These achievements coincided with a continuing revolution in publishing that, by the 1830s, was bringing scientific news to a huge readership. As the Unitarian clergyman William Ellery Channing observed: "Science has now left her retreats . . . her selected company of votaries, and with familiar tone begun the work of instructing the race. . . . Through the press, discoveries and theories, once the monopoly of philosophers, have become the property of the multitudes. . . . Science, once the greatest of distinctions, is becoming popular."[4] Indeed, it was.

The relative ease with which educated Americans succeeded in bringing once-suspect theories in astronomy and geology into harmony with the Bible gave many of them confidence that the same methods would work in the future. In the mid-1840s, a backwoods astronomer in Pennsylvania, Daniel Kirkwood, discovered a "law" governing the axial rotations of the planets in the solar system, based on assuming the truth of Pierre-Simon Laplace's nebular hypothesis. This led some proud fellow countrymen to dub Kirkwood "the American Kepler," in homage to seventeenth-century German astronomer Johannes Kepler, who had famously discovered the three laws of planetary motion. Kirkwood's contribution seemingly encouraged American men of science to embrace Laplace's naturalistic theory of the development of the solar system more readily than their British and continental colleagues, who derived no nationalistic pride from the Yankee's accomplishment.[5]

Additional challenges to religious orthodoxy came with the appearance of such "pseudo-sciences" as phrenology, Mesmerism, and materialistic physiology in the middle third of the nineteenth century. The spread of such "infidel science," as it was frequently denounced, provoked a religious backlash, characterized by a torrent of popular literature on "science and religion," a novel phrase that began appearing with frequency in the period after about 1830.[6]

The School of American Ethnology

In the years before the appearance of Charles Darwin's *On the Origin of Species*, nothing epitomized "infidel science" for American Christians more than the polygenetic theory of human origins. During the 1840s, a decade that saw anthropology emerge as one of the "new and frisky sciences," the editor of the London-based *Ethnological Journal* dubbed the Philadelphia naturalist-physician Samuel George Morton and his disciples "the school of American Ethnologists," a designation proudly adopted by the members themselves.[7] Although Americans had for decades been speculating about the origin of the various human races, this seems to be the first time that international observers identified Americans as constituting any scientific school, this one characterized by a rejection of the singular creation of Adam and Eve (monogenism), as described in the Bible, in favor of multiple creations of the ancestors of the human races (polygenism).[8]

The controversy began in the late 1830s, when Morton published his book called *Crania Americana*, in which he drew on his impressive collection of human skulls to argue for distinctive racial differences in cranial capacity. At the time, Morton apparently still accepted the theologically orthodox "unity of the human species," but by the mid-1840s, he had embraced the heretical polygenist position. In 1850, a year before his death, he avowed his "belief in a plurality of origins for the human species," writing to a friend that he had adopted this position four years earlier—"with some misgivings, not because I doubted the truth of my opinions, but because I feared they would lead to some controversy with the clergy."[9]

By that time, Morton had come under the influence of two vocal supporters, George R. Gliddon, the United States consul at Cairo, who collected Egyptian skulls for him, and Josiah Clark Nott, a physician from Mobile, Alabama, both of whom possessed a zeal to free anthropology, like astronomy and geology before it, from the shackles of Scripture. Together in 1854, Nott and Gliddon brought out a 738-page treatise, *Types of Mankind*, dedicated to the late Dr. Morton and his polygenist views. Included in the volume was a contribution from Louis Agassiz, now the leading naturalist in America, who had embraced polygenetic theory a few years after emigrating from Switzerland in the mid-1840s. Gliddon and Nott might easily be dismissed as "charlatans," but the same could hardly be said of Agassiz or even of Morton, whose very presence in the polygenetic camp made it difficult to dismiss their anthropology as pseudoscience.[10]

Leading the opposition to polygenism was South Carolinian John Bachman, a Lutheran pastor and naturalist. Bachman was well known in the scientific community for having collaborated with his close friend John James Audubon on *The Viviparous Quadrupeds of North America* (1845–1848), a magisterial study of American mammals. Despite his ministerial credentials, he engaged the polygenists primarily on scientific rather than biblical grounds, arguing, on the basis of his own experiments, for the sterility of hybrids, which would make it virtually impossible for different species of humans to procreate. The Charleston Literary Club, which had hosted Agassiz in 1847, served as the forum for debating the unity of the races. The Reverend Thomas Smyth, pastor of the Second Presbyterian Church of Charleston, joined Bachman in defending monogenism. The remainder of the city's intelligentsia, including its impressive circle of naturalists and physicians, supported Agassiz. One of these, an Anglican minister on the faculty of the College of Charleston, said that he "would not give a fig for the benevolent Dr.'s [Bachman's] recommendation of anything except, perhaps, a partridge or a rat. I don't trust his judgment respecting the genus homo." The editors of the *Charleston Medical Journal* gave Morton space to cut "old Bachman . . . into sausage meat," as Nott described it. Bachman had riled the Alabama physician in his review of *Types of Mankind* by noting that "the world of Science has never admitted . . . [Nott and his polygenist colleagues] into their ranks as Naturalists. Their names are utterly unknown among them—not one . . . has ever described a single animal." Apparently, his anger caused Bachman to forget that the leading zoologist in the country, Agassiz, was also a polygenist.[11]

Some contemporary observers and a few historians have suggested that the polygenist school dominated midcentury scientific thinking about race in America.[12] An inventory of known opinions indicates the opposite. Despite the South's overwhelming preference for the biblical story of God's curse on Noah's son Ham, one of the strongest centers of support for polygenism appears to have been what Lester D. Stephens has identified as the "Charleston circle of naturalists," a group that included at least seven advocates of multiple human origins, the majority being physicians. The College of Charleston naturalist Lewis Reeve Gibbes notably remained above the fray.[13] Contributors and subscribers to *Types of Mankind* provide another source of likely polygenists. According to Stephens, "two relatively unknown physicians," William Usher and Henry S. Patterson, published relatively brief essays in the book, along with major contributions from Nott, Gliddon, and Agassiz. Listed among the more than 350 subscribers to the volume were the names of the northern naturalists Timothy Conrad, Samuel S. Haldeman, John Eatton LeConte, Joseph Leidy, J. Aiken Meigs, and Charles Pickering; the Charleston naturalists Francis Simmons Holmes, John Edwards Holbrook, and Edmund Ravenel; and Michael Tuomey, the state geologist of Alabama.[14] Not all subscribers, however, fully embraced polygenism. Leidy, despite repeated pressure from Nott, refused to declare his position; Pickering, though a friend of Morton's, avoided stating his views publicly.[15] From the work of David Livingstone, we know that Harvard geologist and paleontologist Nathaniel Southgate Shaler, a student of Agassiz's, embraced polygenism.[16]

In contrast, most of the country's scientific elite, with the exception of Morton and Agassiz, rejected polygenism. Among the monogenists were the country's foremost botanist, Asa Gray; its most distinguished geologist, James Dwight Dana; its most prominent physicist, Joseph Henry, who presided over the Smithsonian Institution; Alexander Dallas Bache, superintendent of the US Coast Survey; and such influential anthropologists as Henry Schoolcraft, Albert Gallatin, and Lewis Henry Morgan.[17] Other well-known opponents of plural human origins included Princeton naturalists Arnold Guyot and John Torrey; Amherst cleric-geologist Edward Hitchcock; Moses Ashley Curtis, the foremost authority on the botany of the southern Appalachian Mountains; and the physicians Samuel Forrey, Bennet Dowler, and James Lawrence Cabell.[18] Two of Canada's most influential men of science, Montreal's John William Dawson and Toronto's Daniel Wilson, also condemned polygenism, which the latter termed "the American point of view."[19] Perhaps the most surprising opponent of polygenism was John William Draper, a leading American positivist and author of the virulently anti-Catholic *History of the Conflict between Religion and Science* (1874), who on scientific grounds rejected the notion "that there have been as many separate creations of man as there are races which can be distinguished from each other."[20]

Several years ago, G. Blair Nelson observed that "many of the country's most influential men of science rejected polygenism but remained silent on the issue because they had no good scientific alternative to offer and few convincing answers to the American school's arguments." This seems to explain the behavior of Othniel C. Marsh, Yale's professor of vertebrate paleontology who later served as president of the National Academy of Sciences. In what appears to be his only published opinion on the controversial subject of polygenism, Marsh concluded a note on the excavation of an ancient burial mound with the hope that additional archaeological investigations would "doubtless also help to solve the question of the antiquity of man on this continent, and, perhaps, that more important one of the unity of the human race."[21]

Evolution

In the American scientific community, the rapid acceptance of Charles Darwin's contentious but monogenetic theory of human origins after the appearance of his books *On the Origin of Species* (1859) and *The Descent of Man* (1871) deflated the polygenist balloon. American naturalists played a significant role in providing evidence of evolution. The trans-Mississippi West proved to be a vast natural laboratory for studying the history of life on Earth. The most compelling empirical evidence in support of evolution came to light in the early 1870s, when Othniel C. Marsh uncovered the fossil remains of toothed birds in Cretaceous rocks from Kansas. By the 1870s, the large number of fossils discovered in the West provided compelling empirical evidence in support of evolution, serving in the words of Marsh as "the stepping-stones by which the evolutionist of to-day leads the doubting brother across the shallow remnant of the gulf once thought impassable." These findings, raved British zoologist Thomas H. Huxley, elevated Charles Darwin's speculations about missing links "from the region of hypothesis to that of demonstrable fact."[22]

In 1876, Marsh won even greater fame for assembling a series of equine fossils that remarkably documented the evolution of the horse in North America. This evidence prompted Huxley, then lecturing in the United States, to announce that "the doctrine of evolution, at the present time, rests upon exactly as secure a foundation as the Copernican theory of the motions of the heavenly bodies did at the time of its promulgation." A few years later, Darwin himself congratulated Marsh for providing "the best support to the theory of evolution, which has appeared within the last 20 years."[23]

In 1872, fewer than thirteen years after the appearance of Darwin's *On the Origin of Species*, the Philadelphia-based paleontologist Edward Drinker Cope, who competed ferociously with Marsh for control of the Western fossil fields, observed that "the modern theory of evolution has been spread everywhere with unexampled rapidity, thanks to our means of printing and transportation. It has met with remarkably rapid acceptance by those best qualified to judge of its merits, viz., the zoologists and botanists." Just before his death in 1873, even Louis Agassiz, the arch-critic of developmental theories, conceded that the idea of organic development had won "universal acceptance." The nearly unanimous approval of evolution within the American scientific community was spotlighted in 1880, when the editor of the relatively liberal Congregational *Independent*, after surveying the teaching of evolution in the colleges of the North, challenged a rival religious weekly, the conservative Presbyterian *Observer*, to name "three working naturalists of repute in the United States—or two (it can find one in Canada)—that is not an evolutionist." Besides Canadian John William Dawson of McGill, the *Independent* could discover only one other self-identified non-evolutionist: Princeton's Arnold Guyot, who minimized the number of divine interventions.[24]

Allegedly distinctive American ideas played little role on the international scientific stage until the outbreak of controversies over evolution in the late nineteenth century. As early as 1876, the entomologist Alpheus Spring Packard Jr. referred to "an original and distinctively American school of evolutionists," which he subsequently dubbed "Neolamarckism." This loose coalition of American naturalists sought primarily to identify the natural, external causes of organic variations, especially the influence of climate. As Packard noted, "The influence of climate on variation has been studied to especial advantage in North America, owing to its great extent, and to the fact that its territory ranges from the polar to the tropical regions, and from the Atlantic to the Pacific." Neo-Lamarckians tended to downplay the role of natural selection, typically assigning it a secondary function in weeding out unfit organisms. Contrary to popular opinion, most of them assigned a limited role to the heritability of acquired characters, which even Darwin accepted. American neo-Lamarckians, who embraced naturalism as enthusiastically as did the Darwinists, conformed to no distinctive religious profile.[25]

Anti-evolution

The infamous Scopes trial in 1925 in Dayton, Tennessee, in which a young high school teacher was convicted of violating the recently enacted state law banning the teaching of human evolution, pitted a leading Democratic politician, William Jennings Bryan (for the prosecution), against the prominent agnostic lawyer Clarence Darrow. According to one Tennessee newspaper: "In Constantinople and far Japan, in Paris and London and Budapest, here and there and everywhere, at home and abroad, in pagan and in Christian lands, where controversialists gather, or men discuss their faith, to speak the name of Dayton is to drop a bomb, to hurl a hand grenade, to blow the air of peace to flinders." One American scholar calculated that the trial was discussed by "some 2310 daily newspapers in this country, some 13,267 weeklies, about 3613 monthlies, no less than 392 quarterlies, with perhaps another five hundred including bi-monthlies and semi-monthlies, tri-weeklies and odd types." His search turned up "no periodical of any sort, agricultural or trade as well, which has ignored the subject." For the first time in history, radio transmitted a science-and-religion debate to the people. A new day in the popularization of science and religion had commenced.[26]

As readers around the world tracked events in Tennessee, many concluded that conservative Christians in the United States, particularly in the South, had little use for modern science. This led to much finger pointing, jesting, and shaming of American culture—especially, it appears, in the English-speaking world. British commentators could scarcely contain their glee at what had befallen their benighted American cousins. After surveying some forty-four British periodicals, the historian Dean Rapp observed that the trial provided "an outlet for their postwar resentments of the accelerating Americanization of British popular culture, and the growing wealth and power of the United States." Implicitly or explicitly, they assured themselves, in the words of the *Saturday Review*, a London weekly, that "such things can only happen in America." The *Manchester Guardian* snidely observed that "only by going to the United States can we escape from the modern world."[27]

The day after the trial opened, *Nature*, the leading science journal in Britain, published a sampler of how thirty-two distinguished scientists, educators, and divines viewed the bizarre goings-on across the Atlantic, which the editors referred to as "intellectual terrorism." Sir Arthur Shipley, master of Christ's College, Cambridge, dismissed "the average American of the Middle and Southern States" as "a very naïve mammal," far more "childish" than the citizens of "older and more mature countries." This elite anatomist focused his disrespect on "the farmers and the Methodist and Baptist pastors" of the South. The Right Reverend E. W. Barnes, lord bishop of Birmingham, deplored "the ignorant fanaticism which has led to the proscription of evolution in certain Western States of America," confusing the geographical focus of the furor and failing to note that the campaign targeted the teaching of human evolution, not organic evolution generally. William Johnston Sollas, professor of geology at the University of Oxford, condescendingly employed language "that will be readily understood by our Puritan friends" to tell them that "all zoologists and botanists are agreed that the creation of species, including man, proceeds or has proceeded by way of evolution." Sir Arthur Keith, Hunterian professor and conservator of the Museum at the Royal College of Surgeons of England, benevolently quoted the words of Jesus regarding those who crucified him: "Father, forgive them for they know not what they do." Francis Bather, keeper of the Department of Geology at the British Museum, dismissed the Tennessee law as a "medieval gesture." Others deployed such terms of "preposterous" and "astonishing" and described American anti-evolutionists as "obscurantist heresy hunters." J. McKeen Cattell, editor of the American journal *Science*, fired off a response, pointing out the limitations of the anti-evolution law (to Tennessee, to state-supported institutions, and to human evolution), observing that "there is a larger proportion of "Fundamentalists" in every European nation than in the United States."[28]

A number of other prominent Britons chimed in with their opinions of the trial. "With us the question of teaching Darwinism in the schools has never arisen, and it seems incredible to us that it ever could arise," declared David Lloyd George, former prime minister of Great Britain. The writer H. G. Wells, author of the best-selling two-volume *Outline of History* (1920), took the opportunity to advertise his work in the *New York Times Book Review*, advising that "Really to Understand Evolution—Read the *Outline of History*" and noting that "one thing at least has resulted from the Tennessee indictment of Mr. Scopes. Today, more than ever, it is inexcusable in an otherwise educated person, not to understand what Evolution really means." In a much-covered debate on the eve of the trial, the British playwright George Bernard Shaw, who that year won the Nobel Prize in Literature, dismissed Bryan as "a man with . . . no discoverable brains of any kind, and one of those men which America alone seems to produce." He diagnosed fundamentalism as "infantilism, in the pathological sense"; restricting the teaching of evolution constituted "the dogma of a blockhead." Unlike many of his countrymen, Shaw predicted that American fundamentalism, like American cinema, would eventually invade the United Kingdom.[29] And it did. In 1927, Arthur Keith, then president of the British Association for the Advancement of Science, devoted his presidential address to "Darwin's Theory of Man's Descent as It Stands

To-Day," in which he laid out the compelling evidence for human evolution. Newspapers took his message to what one account termed "the plain people," reawakening the long dormant controversy over the subject. Annoyed by the controversy, Keith charged his critics with showing that "Daytonism is alive and vigorous in England. There are still, I am surprised to learn, many men and women who are convinced there is only one reliable account of man's origin—that given in the Book of Genesis." His surprise, noted one commentator, stemmed from the fact "that the educated Englishman, accustomed more or less to ignoring the opinions of the common people of his country, has insisted on comparing the beliefs of England's intelligentsia with those of American's masses."[30]

Elsewhere in the British Empire, the "Tennessee monkey trial" also attracted considerable comment. Canadians largely avoided the fundamentalist-modernist schism that divided Protestants to their south—and so were able to view the Scopes trial with bemusement. An editorial in one Manitoba paper depicted the trial as an example of "American foolishment," during which young Scopes "was exposed to the tender mercies of as proficient a gang of professional poseurs, wind-jammers, self-advertisers and self-exploiters as the United States can boast."[31] Perhaps the most notable exception was T. T. Shields, a biblical inerrantist who pastored Toronto's prosperous Jarvis Street Baptist Church and edited *The Gospel Witness*, and who had emerged as one of the leading fundamentalists in all of North America. Shields actively participated in the World's Christian Fundamentals Association, which assisted with the prosecution of Scopes and arranged for Bryan to participate. Shields in time took credit for helping to bring Bryan to Dayton, saying on one occasion: "I count it an honour to have known him; I shared with other brethren of the World's Christian Fundamentals Conference when it met in Memphis, in sending him a telegram asking him to assume the responsibility of assisting in the case against Evolution in Tennessee." At times, Shields used his church to host creation-evolution debates. On hearing a fellow Canadian Baptist say that "any man who refuses to believe the doctrine of evolution makes himself ridiculous and puts himself without the pale of educated men," Shields responded without a trace of shame: "Well, I frankly admit that I must be ridiculous and I am without the pale of educated men if that be true."[32]

A second major Canadian participant in the anti-evolution controversies of the 1920s was George McCready Price, dubbed "the principal scientific authority of the Fundamentalists" by the editor of *Science*. Born in rural New

Brunswick, Price as a youth converted with his widowed mother to a small apocalyptic sect known as the Seventh-day Adventists, founded in the mid-nineteenth century by a charismatic prophet named Ellen G. White. Because they worshiped on Saturday as a memorial of a six-day creation, Adventists adamantly opposed any scientific theory that proposed interpreting the days of creation symbolically. They also viewed Noah's flood as a worldwide catastrophe that had buried the fossils and reshaped the earth's surface. In the early 1890s, Price enrolled for two years in the "classical course" at Battle Creek College, an Adventist school in Michigan. Later, he matriculated in a one-year teacher-training course at the Provincial Normal School of New Brunswick (now the University of New Brunswick), where he took "some elementary courses in some of the natural sciences, including mineralogy." That was the extent of his formal training in science.[33]

Over the years, Price developed an avid interest in geology. In 1902, while working as a school teacher in the Maritimes, he wrote the first of numerous books expounding on White's creationist insights, *Outlines of Modern Christianity and Modern Science*, which he liked to call "the first Fundamentalist book."[34] From 1924 to 1928, Price taught at an Adventist college near London, where he faithfully attended the meetings of the Victoria Institute, the foremost anti-Darwinist organization in Great Britain. To his surprise, members of the institute expressed little sympathy for what they regarded as his peculiarly American brand of anti-evolutionism. The editor of its journal took exception to Price's attempt to foment "a new crusade against Evolution" in the United Kingdom; another member condemned Price's effort "to drive a wedge between Christians and scientists" in Britain, as he had done across the Atlantic.[35]

On the eve of the Scopes trial, Bryan, who had met Price and occasionally corresponded with him, invited the Canadian to join him in Dayton as an expert witness for the prosecution, addressing him as "one of the outstanding scientists who reject evolution." Because of his absence from North America, Price declined Bryan's invitation, which prevented him from hearing Darrow's scathing dismissal of him as "a mountebank and a pretender and not a geologist at all."[36] Despite his prominence in the antievolution movement, Price, who spent many of his adult years in the United States, seems to have attracted little attention in his native land.

In the British Caribbean, Jamaica's newspaper of record, the *Kingston Daily Gleaner*, kept its readers fully informed about the Scopes trial throughout the summer of 1925. The city's popular Bournemouth Baths repeatedly ran an advertisement promoting their establishment while poking fun at the United States. The text of this ran:

> **THE SCOPE EVOLUTION TRIAL!!!**
> Our big neighbor—the U.S.A., has been having a real good time with the theories of Evolution—relativity—"cubism"—Fundamentalism and other "isms"—but that's ALL THEIR OWN LITTLE PARTY, and we're merely SPECTATORS. We pass no opinion on the "pros and cons" of Evolution in THIS test case—but:
> **We believe in Evolution—Yes—we surely do!**
> **We know that constant Bathing at Bournemouth**
> **Causes Evolution!!**
> Because the SUBSEQUENT healthy condition of yourself and your spirits is a NATURAL EVOLUTION from the SLUGGISH, WEARY "ALL-IN" FEELING that you experienced PRIOR to the CONSTANT SEA BATHING EXERCISE.
> You won't "Evolute" into a fish after constant Sea Bathing
> But if your friends call you a "poor fish" for not bathing Bournemouth often!!—Then it's all your own fault!![37]

Not surprisingly, interest in the Scopes trial extended all the way to the South Pacific, especially to Australia and New Zealand. In 1902, a telegraph cable finally connected North America and Australia, making it possible to transmit daily accounts of the trial twenty-three years later. "We are told," Bryan reported at the close of the trial, "that more words have been sent across the ocean by cable to Europe and Australia about this trial than has ever been sent by cable in regard to anything else doing in the United States." Western Union alone brought twenty-two key operators to Dayton to assist with the unprecedented traffic.[38] Beginning shortly before the trial opened, the Australian Cable Service took the lead in furnishing the Dominion's newspapers, from Perth to Wagga Wagga, with day-to-day reporting of what had transpired in Dayton, Tennessee.[39] The *Advocate*, published for the Catholic Archdiocese of Melbourne, summed up the level of excitement this way: "Even in far-away Melbourne the doings in Dayton are chronicled with a wealth of detail usually only conceded to the coronation of a monarch and a wrestling match at the Stadium."[40] Although most headlines simply described the trial as "interesting" or "a test case," others could not resist catchier phrases such as "Freak Trial," "Great Monkey Trial," "Holy War," "America's New Stunt," "A Bizarre Trial: And So Typically American."[41]

As the last headlines illustrate, the Australian coverage of the trial, like that elsewhere, tended to highlight American idiosyncrasies. This was especially true in editorials. "Naturally the idea that the State had a right of supervision over the morals and thoughts of its citizens could not last," observed the *Newcastle Sun*, "but in America, almost untouched by progress, the Puritan domination persisted long after the Restoration had swept it away in England." Comparing Scopes' ordeal with the trial of Galileo and the burning of witches, the editorial concluded: "No other country, at this stage of the world's progress, could possibly have staged such a trial." The weekly Perth *Truth* expressed its surprise "to find, not only existing but persisting, in a country like the U.S.A., which boasts the most complete and up-to-date educational system in the world, conditions which were common in the dark days of the Spanish Inquisition, and further back in the days of Nero." Perhaps, speculated the editorialist, "it is just an offshoot of the general prohibition mania" in the "Yewnited States." The venerable *Adelaide Register* described the trial as an "exquisite piece of folly" and raised the possibility of "mass hysteria." Pointing to American exceptionalism "on subjects of this kind," the paper noted that "in almost every affair relating to the material facts of life, the Americans are amazingly practical and progressive; but, where their emotions are concerned, many millions of them are prone to kinds of surprising excesses, and cranks of every variety may count upon a numerous and eager following." One of the rare newspapers to note that horror of evolution was not unique to the United States was the *Brisbane Worker* affiliated with the Australian Labor Party, which observed that many antievolutionists could "be found, not only in America, but in Australia."[42]

By coincidence, the Scopes trial occurred just after the Australian physician Raymond Dart, then working in South Africa, announced his discovery in Taung of a child's fossilized skull, the first of its kind found in Africa. Dart named it *Australopithecus africanus*—or "the Man-Ape of South Africa"—and identified it as an extinct ancient ancestor of humans. In the controversy that ensued, another Australian, Grafton Elliot Smith, who had gone from Sydney to the chair of anatomy at the University College London, defended his fellow countryman against the influential anatomist Sir Arthur Keith. "Both Dr. Dart and Professor Elliott Smith are Australians," bragged the *Newcastle Sun* in a piece that compared Australia and Tennessee: "The great majority of people are quite indifferent as to who their ancestors were, but there is a great deal of satisfaction in knowing that the Australian Universities can turn out men like Smith and Dart, who are able to take an honored place in the world's most advanced scientific circles."[43]

As in many other places, the New Zealand press covered the Scopes trial with voyeuristic pleasure. The *New Zealand Free Lance*, a popular illustrated weekly, described the event as "the most amazing trial held since the days of the Spanish Inquisition, or, say, the witchcraft 'smelling out' era in Massachusetts." The prosecution of Scopes showed "America's freak laws at their zenith of silliness" and provided the world with "the joke of the decade." The day after the trial began, the *New Zealand Herald* ran a front-page story captioned "A Crusade of Darkness: Blind Fanaticism." Linking the trial to religious dogmatism and Southern racism, the newspaper found it "hard to take the anti-evolution movement seriously." Such rhetoric established the public image of antievolutionists as benighted fools, indigenous to the southern United States. As a correspondent to the *Otago Daily Times* proudly declared, he lived in "NEW ZEALAND, NOT TEXAS."[44]

Such complacency was shattered in the late 1920s, when the evolution controversy struck home in New Zealand. In 1928, the Education Department added a clause to the *Syllabus of Instruction for Public Schools* suggesting that in the higher classes the pupils should gain some definite ideas of the principle of evolution. The department recommended a couple of history texts that discussed the evolutionary origins of humans. These changes outraged New Zealand's small but vocal community of fundamentalists. In repeated letters to the editor of the *New Zealand Herald* and in a pamphlet issued by Auckland's fundamentalist Bible Training Institute, William H. Pettit, a physician on staff who taught a course on "The Bible and Science," dismissed evolution as an unproven hypothesis, used by the enemies of Christianity as "the chief weapon of attack upon the Bible." To bolster his argument, he cited a number of scientists who had not accepted evolution, including the deceased Canadian geologist Dawson and the Seventh-day Adventist creationist Price.[45] One unimpressed reader, a self-identified rationalist, noted that Pettit's creationists were all "dead men, nonentities and—staunch evolutionists." He surmised that the doctor, "the chief scientific pillar of the fundamentalists of the Dominion in their warfare against Evolution," had "forgotten the scientific smatterings that form part of the medico's course, and in his zeal for his Biblical preconceptions . . . [had] uncritically absorbed the special pleading of his Roman Catholic and Seventh Day Adventist [*sic*] 'authorities.'" In his eagerness to expose Pettit's scientific ignorance, the critic revealed his own: he dismissed the influential American naturalist and evolutionist Joseph LeConte as "a minor French philosopher."[46]

Meanwhile, on South Island New Zealand, the Presbyterian Reverend Philadelphus Bain Fraser, who had once hoped to entice William Jennings Bryan to New Zealand, bombarded the minister of education, the governor-general, members of the legislature, and other officials with angry letters objecting to the exposure of children to "evolutionary propaganda." Eager now to distance himself from American antievolutionists, who had recently won approval for an antievolution referendum in Arkansas, he denounced such actions as "the worst form in which to present complicated religious and quasi-religious issues for settlement by the people," explaining that *he* simply desired fairness. "Why," he asked in the *Otago Daily Times*, "should the dogma of evolutionism be projected into the schools and the dogma of Christianity be excluded?" Fraser's outburst, subsequently published in pamphlet form as *Evolution and Ape-Manism in Schools*, prompted one reader to advocate the teaching of evolution as an antidote to "religious superstition." Yet local Christians responded readily to Fraser's crusade. At a four-hundred-strong meeting in Christchurch, almost everyone resolved to oppose the teaching of evolution in schools. Embarrassed by the "widespread misunderstanding" created by the new syllabus, the Ministry of Education sought to clarify its position by backpedaling as fast as it could. Professing its commitment to a divine Creator, the ministry urged teachers to avoid bringing any unpleasant controversy into the classroom, thus, in effect, handing the Kiwi antievolutionists a victory.[47]

Creationism Goes Global

For several decades following the 1920s, debates over evolution—and science and religion generally—faded from the world's headlines. This changed dramatically in the early 1980s, when two American states, Arkansas and Louisiana, passed so-called balanced-treatment acts that required teaching "creation science" whenever "evolution science" was introduced. In a headline-attracting trial in 1981–1982 in Little Rock, nicknamed Scopes 2, a federal judge ruled the Arkansas law unconstitutional. A short time later, another federal court decreed similarly without a trial, a judgment upheld by the US Supreme Court in 1987.[48]

Simultaneously with all of this legislative and legal activity, American advocates of "creation science"—based largely on Price's flood geology—began successfully proselytizing around the world for this distinctive version of creationism. For a few years, American scientists tried to reassure other nations that they had nothing to fear from this American threat. In 1986, the distinguished American paleontologist and anti-creationist Stephen Jay Gould, then visiting Auckland, assured New Zealanders that they had little to fear from scientific creationism. Because the movement was so "peculiarly American," he thought it stood little chance of "catching on overseas." Fourteen years later, he was still assuring listeners that creationism was not contagious. "As insidious as it may seem, at least it's not a worldwide movement," he asserted: "I hope everyone realizes the extent to which this is a local, indigenous, American bizarrity." Gould's colleague Richard C. Lewontin made the same point, as did Edward O. Wilson, another Harvard biologist.[49]

Although Gould and his Harvard associates remained oblivious to it, the worldwide growth of creationism since the early 1980s had already proved these observations and predictions utterly wrong. Antievolutionism had already become a global phenomenon—and was rapidly spreading from conservative Christianity to Judaism and Islam. Not surprisingly, a common—though not insurmountable—impediment to its progress was the attached label "Made in America."[50]

No country outside the United States gave creationism a warmer reception than Australia, which spawned an internationally successful creationist ministry and at times even welcomed creation science into the classrooms of state-supported schools. As late as 1984, one of the best-informed students of Australian fundamentalism predicted that "because of the different national traditions and educational systems, the [creationist] controversy is not likely to become as intense in Australia as in [the] USA." But already young-earth creationists, led by schoolmaster Ken Ham and physician Carl Wieland, had organized a Creation Science Foundation (CSF), which, from its headquarters in Brisbane, in Queensland, quickly became the epicenter of antievolutionism in the South Pacific. Queensland offered a unique opportunity for Australian creationists, in part because Protestantism in the state had become heavily Americanized since World War II and because "creation" already appeared in the official science syllabus for public schools. "Unshackled by constitutional restraints" on the teaching of religion in public schools, as one critic put it, Australian creationists were "not as coy as their American counterparts in declaring their evangelical purposes." This situation "led in Australia to a cruder, more forthright, aggressive, and less subtle form of creation science than is the case in the USA."[51]

Creationist critics, ever eager to discredit creationism in the eyes of the public, rarely passed up an opportunity to stress its American origins and its unsuitability for Australia, which, one insisted, was not merely "the USA with kangaroos." Australian intellectuals dismissed creationism as "an American anti-intellectual anachronism" and "a reversion to America's Deep South of the '20s," with one commentator describing it as a "recent American import" brought over by visiting speakers from the United States. Another suspected the US government of secretly funding the CSF. A popular journalist declared that he would "rather have US military bases in this country than see Queensland become a base for this sort of hillbilly nonsense."[52]

Because of Australian sensibilities about American cultural imperialism, creation scientists from the beginning sought to distance themselves from their American roots: modifying US texts for local consumption, marketing the magazine *Ex Nihilo* as "Australia's own Creation-Science magazine," and insisting that the CSF was "utterly distinct from overseas bodies," particularly in its emphasis on the biblical basis for creationism. With less fanfare, similar developments occurred in New Zealand, where in 1992 creationists set up a "NZ arm" of the CSF, called Creation Science (NZ).[53]

Most cosmopolitan Canadians, according to one observer, liked to believe that creationism stopped at the US border. One explanation for this is that Canadian newspapers traditionally have not paid much attention to the topic. And when it does appear, notes anthropologist John Barker, "its proponents are presented as, at best, quixotic crusaders but often as 'American-style' religious fanatics bent on imposing their beliefs upon Canadian society. . . . There is a pervasive strain of anti-American attitudes especially among English-speaking Canadians which mixes in complex ways with an equally strong attraction to American culture." Needless to say, Canadian creationists have tried to establish their own creationist culture.[54]

During the late twentieth century, creationism began to make a splash among evangelical Christians in Great Britain. Few nonbelievers gave it a thought before the early twenty-first century, when the inroads of American proselytizing became apparent. This prompted one London biologist to warn "US-Style Creationism Spreads to Europe." One public-opinion poll revealed that only 48 percent of Britons believed that the theory of evolution "best described their view of the origin and development of life." A sixth-form biology teacher in London reported that "the vast majority" of her brightest students rejected evolution. "It's a bit like the southern states of America," she noted.[55]

As this last observation suggests, a major impediment to the propagation of young-earth creationism in the UK was its "Made in America" label. Critics of the British creationists accused them of "injecting an alien element into British evangelicalism" and importing "the shibboleths of American fundamentalism." Some opponents alleged ties to "right-wing North American politics." Because of the widespread prejudice against all things American, British creationists tended to distance themselves from the American movement, stressing the indigenous roots of creationism in the local evangelical reformation and, where possible, quoting British rather than American authorities. At least one young-earther alleged that British creationists who denied their intellectual debt to Americans were reacting culturally, not historically. He invoked the memory of Dwight L. Moody's revivals in the 1860s and Billy Graham's campaigns in the 1940s to answer the question "Can any good thing come out of America?" with a resounding "Yes."[56]

Beyond the United Kingdom, creationism in Europe caused few ripples but ran deeper than many Europeans suspected—or were willing to admit. From time to time, warnings about the spread of creationism appeared. In 2000, for example, the British magazine *New Scientist* devoted a cover story to warning the public that it was time to "Start Worrying Now," because "From Kansas to Korea, Creationism Is Flooding the Earth." Although the unexpected development seemed almost "beyond belief," the magazine described creationism as "mutating and spreading" around the world, "even linking up with like-minded people in the Muslim world." Five years later, a Dutch science writer published an article in *Science* provocatively titled "Is Holland Becoming the Kansas of Europe?" Contrary to almost all expectations, geographical, theological, and political barriers had utterly failed to contain creationism.[57]

Ironically, some of the New Atheist critics of creationism have themselves come under attack for trafficking in "a set of distinctly American presumptions about the relationship between religion and science." This, alleges one accuser, constitutes nothing less than "Americanized atheism." The "Made in America" critique can cut both ways.[58]

Notes

1. David N. Livingstone, "Science, Region, and Religion: The Reception of Darwinism in Princeton, Belfast, and Edinburgh," in *Disseminating Darwinism: The Role of Place, Race, Religion, and Gender*, ed. Ronald L. Numbers and John Stenhouse (Cambridge: Cambridge University Press, 1999), 7–38; David N. Livingstone, *Putting Science in Its Place: Geographies of Scientific Knowledge* (Chicago: University of Chicago Press,

2003); David N. Livingstone, *Dealing with Darwin: Place, Politics and Rhetoric in Religious Engagements with Evolution* (Baltimore: Johns Hopkins University Press, 2014).

2. Oliver Wendell Holmes et al., "Report of the Committee on Literature," *Transactions of the American Medical Association* 1 (1848): 286–87; [Sydney Smith], "Review of *Statistical Annals of the United States of America*, by Adam Seybert," *Edinburgh Review* 33 (1820): 79.

3. Benjamin Silliman, *American Contributions to Chemistry: An Address Delivered on the Occasion of the Celebration of the Centennial of Chemistry, at Northumberland, Pa., August 1, 1874* (Philadelphia: Collins, 1874), 66–69; [John] Swinton, "The Recent Congress at Newport—Resume of Results," *New York Times*, August 25, 1860, 1. See also Robert V. Bruce, *The Launching of Modern American Science, 1846–1876* (New York: Alfred A. Knopf, 1987).

4. William E. Channing, *The Present Age: An Address Delivered Before the Mercantile Library Company of Philadelphia, May 11, 1841* (Philadelphia: J. Crissy, 1841), 10.

5. Ronald L. Numbers, "The American Kepler: Daniel Kirkwood and His Analogy," *Journal for the History of Astronomy* 4 (1973): 13–21; Ronald L. Numbers, *Creation by Natural Law: Laplace's Nebular Hypothesis in American Thought* (Seattle: University of Washington Press, 1977), 41–54.

6. On the recently coined term "pseudoscience," see Daniel P. Thurs and Ronald L. Numbers, "Science, Pseudoscience, and Science Falsely So-Called," in *Wrestling with Nature: From Omens to Science*, ed. Peter Harrison, Ronald L. Numbers, and Michael H. Shank (Chicago: University of Chicago Press, 2011), 281–305. On "infidel science," see G. Blair Nelson, "Infidel Science! Polygenism in the Mid-Nineteenth-Century American Weekly Religious Press," unpublished PhD dissertation, University of Wisconsin–Madison, 2014.

7. Monte Harrell Hampton, *Storm of Words: Science, Religion, and Evolution in the Civil War Era* (Tuscaloosa: University of Alabama Press, 2014), 63 (frisky), quoting the Southern Presbyterian George Howe; [Luke Burke], "Progress of Ethnology in the United States," *Ethnological Journal* 4 (September 1848): 169–85, quote on 170 and 173. On the adoption of the appellation by members of the American School, see J. C. Nott and Geo. R. Gliddon, *Types of Mankind; or, Ethnological Researches, Based upon the Ancient Monuments, Paintings Sculptures, and Crania of Races* (Philadelphia: Lippincott, Grambo and Company, 1854), 87. On the voluminous history of polygenism in America, see Edward Lurie, "Louis Agassiz and the Races of Man," *Isis* 45 (1954): 227–42; Edward Lurie, *Louis Agassiz: A Life in Science* (Chicago: University of Chicago Press, 1960), 256–65; William Stanton, *The Leopard's Spots: Scientific Attitudes toward Race in America, 1815–59* (Chicago: University of Chicago Press, 1960); George M. Fredrickson, *The Black Image in the White Mind: The Debate on Afro-American Character and Destiny, 1817–1914* (New York: Harper and Row, 1971), esp. ch. 3, "Science, Polygenesis, and the Proslavery Argument"; Robert E. Bieder, *Science Encounters the Indian, 1820–1880: The Early Years of American Ethnology* (Norman: University of Oklahoma Press, 1982), 82; Reginald Horsman, *Josiah Nott of Mobile: Southerner, Physician, and Racial Theorist* (Baton Rouge: Louisiana State University Press, 1987); David N. Livingstone, "The Preadamite Theory and the Marriage of Science and Religion," *Transactions of the American Philosophical Society* 82 (1992): 1–78; Lester D.

Stephens, *Science, Race, and Religion in the American South: John Bachman and the Charleston Circle of Naturalists, 1815–1895* (Chapel Hill: University of North Carolina Press, 2000); Bruce Dain, *A Hideous Monster of the Mind: American Race Theory in the Early Republic* (Cambridge, MA: Harvard University Press, 2002); G. Blair Nelson, "'Men before Adam!' American Debates over the Unity and Antiquity of Humanity," in *When Science and Christianity Meet*, ed. David C. Lindberg and Ronald L. Numbers (Chicago: University of Chicago Press, 2003), 161–81; Michael O'Brien, *Conjectures of Order: Intellectual Life and the American South, 1810–1860*, 2 vols. (Chapel Hill: University of North Carolina Press, 2004), vol. 1, ch. 5; David N. Livingstone, *Adam's Ancestors: Race, Religion and the Politics of Human Origins* (Baltimore: Johns Hopkins University Press, 2008), 173–86.

8. See Nelson, "Infidel Science!"

9. Quoted in Bieder, *Science Encounters the Indian*, 82.

10. Nott and Gliddon, *Types of Mankind*, 87; review of *More Worlds than One: The Creed of the Philosopher and the Hope of the Christian* by David Brewster, *Biblical Repertory and Princeton Review* 26 (1854): 726 (charlatans); [Charles Hodge], "The Unity of Mankind," *Biblical Repertory and Princeton Review* 31 (1859): 103–49, quote on 144 (Agassiz); "Agassiz on Provinces of Creation, and the Unity of the Race," *Biblical Repertory and Princeton Review* 41 (1869): 5–33, quote on 5. I have borrowed parts of this paragraph from Ronald L. Numbers, *Science and Christianity in Pulpit and Pew* (New York: Oxford University Press, 2007), 106–7.

11. Lester D. Stephens, *Science, Race, and Religion in the American South: John Bachman and the Charleston Circle of Naturalists, 1815–1895* (Chapel Hill: University of North Carolina Press, 2000), 170–73 (literary club, Smyth), 186 (sausage), 207 (world of science). On the Bachman-Audubon connection, see also Jay Shuler, *Had I the Wings: The Friendship of Bachman and Audubon* (Athens: University of Georgia Press, 1995).

12. "Natural History of Man," *United States Magazine and Democratic Review* 26 (1850): 327–45, quote on 328. Among the scholars citing this source as evidence of the popularity of polygenism are Jean H. Baker, *Affairs of Party: The Political Culture of Northern Democrats in the Mid-Nineteenth Century* (Ithaca, NY: Cornell University Press, 1983), 179; Adrian Desmond and James Moore, *Darwin's Sacred Cause: Race, Slavery and the Quest for Human Origins* (London: Penguin, 2009), 213; and Robert Bernasconi, "Racial Science," in *A Companion to the History of American Science*, ed. Georgina M. Montgomery and Mark A. Largent (Malden, MA: Wiley Blackwell, 2016), 502–11, quote on 502. British assessments of the American vogue for polygenism include [Luke Burke], "Progress of Ethnology in the United States," *Ethnological Journal* 4 (September 1848): 169–85, quote on 184; and Charles Darwin to J. D. Hooker, March 26, 1854, in Francis Darwin, ed., *The Life and Letters of Charles Darwin*, 2 vols. (New York: D. Appleton, 1897), 1, quote on 403. Terry A. Barnhart, *Ephraim George Squier and the Development of American Anthropology* (Lincoln: University of Nebraska Press, 2005), 295, warns against taking Burke too literally. James Moore, "Darwin's Progress and the Problem of Slavery," *Progress in Human Geography* 34 (2010): 1–28, quote on 14, quotes Darwin to illustrate how British men of science felt about the alleged enthusiasm for Agassiz's views in America.

13. Thomas Virgil Peterson, *Ham and Japheth: The Mythic World of Whites in the Antebellum South* (Metuchen, NJ: Scarecrow Press, 1978), 25–26; Stephens, *Science,*

Race, and Religion, 126 (L. R. Gibbes), 160 (John McCready), 171–74 (James Moultrie, John Edwards Holbrook, Eli Geddings, Robert W. Gibbes), 182 (D. J. Cain, Francis Peyre Porcher). See also H. Shelton Smith, *In His Image, but . . . : Racism in Southern Religion, 1780–1910* (Durham, NC: Duke University Press, 1972), 152–65, and Stephen R. Haynes, *Noah's Curse: The Biblical Justification of American Slavery* (New York: Oxford University Press, 2002).

14. Stephens, *Science, Race, and Religion*, 195; Nott and Gliddon, *Types of Mankind*, 733–38.

15. Leonard Warren, *Joseph Leidy: The Last Man Who Knew Everything* (New Haven, CT: Yale University Press, 1998), 194. On Pickering, see the conflicting opinions of Barry Alan Joyce, *The Shaping of American Ethnography: The Wilkes Exploring Expedition, 1838–1842* (Lincoln: University of Nebraska Press, 2001), 151–54; Livingstone, *Preadamite Theory*, 20; Desmond and Moore, *Darwin's Sacred Cause*, 220; and William Stanton, *The Great United States Exploring Expedition of 1838–1842* (Berkeley: University of California Press, 1975), 339–47.

16. David N. Livingstone, *Nathaniel Southgate Shaler and the Culture of American Science* (Tuscaloosa: University of Alabama Press, 1987), 124.

17. Joseph Henry to Samuel Foster Haven, April 3, 1856, in Marc Rothenberg, ed., *The Papers of Joseph Henry*, vol. 9 (Canton, MA: Science History Publications, 2002), 334, who identifies Gray, Dana, and the Princeton geographer Arnold Guyot as fellow critics of polygenism; A. Hunter Dupree, *Asa Gray, 1810–1888* (Cambridge, MA: Harvard University Press, 1959), 228–29; William Stanton, *The Leopard's Spots: Scientific Attitudes toward Race in America, 1815–59* (Chicago: University of Chicago Press, 1960), 169–70 (Dana, Bache); Stephen O. Murray, *American Anthropology and Company* (Lincoln: University of Nebraska Press, 2013), 6 (Schoolcraft, Morgan, Gallatin).

18. John Torrey to Joseph Henry, April 20, 1850, in Rothenberg, ed., *Papers of Joseph Henry*, 8:35–37; Edward Hitchcock, *Religious Truth, Illustrated from Science: In Addresses and Sermons on Special Occasions* (Boston: Phillips, Samson, and Company, 1857), 92; Edmund Berkeley and Dorothy Smith Berkeley, *A Yankee Botanist in the Carolinas: The Reverend Moses Ashley Curtis, D. D. (1808–1872)* (Berlin: J. Cramer, 1986), 78–81; Samuel Forry, *The Climate of the United States and Its Endemic Influences* (New York: J. and H. G. Langley, 1842), 23–24; Bennet Dowler, "Review of *Types of Mankind*, ed. J. C. Nott and Geo. R. Gliddon," *New Orleans Medical and Surgical Journal* 11 (1854): 108–13; J. L. Cabell, *The Testimony of Modern Science to the Unity of Mankind* (New York: Robert Carter and Brothers, 1859).

19. John William Dawson, *Archaia; or, Studies of the Cosmogony and Natural History of the Hebrew Scriptures* (Montreal: B. Dawson and Son, 1860), 247–48; D[aniel] W[ilson], review of *Indigenous Races of the Earth*, by J. C. Nott and George R. Gliddon, *Canadian Journal: A Repertory of Industry, Science, and Art* 7 (1857): 208–16, quote on 216. See also Carl Berger, *Science, God, and Nature in Victorian Canada* (Toronto: University of Toronto Press, 1983), 42–46 (Dawson and Wilson).

20. John William Draper, *Human Physiology, Statistical and Dynamical; or, the Conditions and Course of the Life of Man* (New York: Harper and Brothers, 1856), 565–68; Stanton, *Leopard's Spots*, 170. The geologist Alexander Winchell was an apparently rare monogenist pre-Adamite; see Livingstone, *Adam's Ancestors*, 141–55, 186–91. See also Lester D. Stephens, "The Earth and Humans before Adam: The Pre-Adamite Theory of

Georgia Geologist Matthew Fleming Stephenson," *Georgia Historical Quarterly* 100 (2016): 9–35.

21. Nelson, " 'Men before Adam!' " 176; O. C. Marsh, "Description of an Ancient Sepulchral Newark, Ohio," *American Journal of Science* 42 (1866): 1–11, quote on 11.

22. William H. Goetzmann, *Exploration and Empire: The Explorer and the Scientist in the Winning of the American West* (New York: Alfred A. Knopf, 1971), 303 (laboratory); O. C. Marsh, "Introduction and Succession of Vertebrate Life in America," *American Journal of Science* 14 (1877): 337–78, quote on 352. On the rapid acceptance of Darwin's theory of evolution, see Ronald L. Numbers, *Darwinism Comes to America* (Cambridge, MA: Harvard University Press, 1998), 24–48. See also Charles Schuchert and Clara Mae LeVene, *O. C. Marsh: Pioneer in Paleontology* (New Haven, CT: Yale University Press, 1940).

23. Thomas H. Huxley, *American Addresses, with a Lecture on the Study of Biology* (New York: D. Appleton, 1877), 90 (Copernican theory); Charles Darwin to O. C. Marsh, August 31, 1880, quoted in Francis Darwin, ed., *The Life and Letters of Charles Darwin*, 2 vols. (New York: D. Appleton, 1896), 2:417.

24. E. E. Cope, *The Origin of the Fittest: Essays on Evolution* (New York: D. Appleton, 1887), 2, from an article "Evolution and Its Consequences," first published in 1871; Louis Agassiz, "Evolution and the Permanence of Type," *Atlantic Monthly* 33 (1874): 92–101, quote on 95; [William Hayes Ward], "Do Our Colleges Teach Evolution?" *Independent* 31 (December 18, 1879): 14–15; William Hayes Ward, "Evolution and Christianity," *Independent* 32 (January 29, 1880), 4; Ronald L. Numbers, *The Creationists: From Scientific Creationism to Intelligent Design*, expanded ed. (Cambridge, MA: Harvard University Press, 2006), 21–22 (Guyot).

25. A. S. Packard Jr., "A Century's Progress in American Zoology," *American Naturalist* 10 (1876), 591–98, quote on 597; Alpheus S. Packard, *Lamarck, the Founder of Evolution: His Life and Work* (London: Longmans, Green, 1901), 383–84. On the limited membership of this "school" among American naturalists, see Numbers, *Darwinism Comes to America*, 33–36. For representative discussions of Neo-Lamarckism as an American construction, see Edward Pfeifer, "The Genesis of American Neo-Lamarckism," *Isis* 56 (1965): 156–67; Joseph M. Maline, "Cope, Edward Drinker," in *Dictionary of Scientific Biography*, ed. Charles Coulston Gillispie (New York: Charles Scribner's Sons, 1970–1978), 15:91–93; George E. Webb, *The Evolution Controversy in America* (Lexington: University Press of Kentucky, 2002), 23, 46.

26. The *Nashville Tennessean*, quoted in "The End of the Scopes Case," *Literary Digest* 92 (February 5, 1927): 14–15, quote on 15; Howard W. Odum, "The Duel to the Death," *Social Forces* 4 (1925): 189–94, quote on 190; "Science Brings Scopes Trial to Midwest's Ears," *Chicago Daily Tribune*, July 15, 1925, 3. For historical accounts of the Scopes trial, see Numbers, *Darwinism Comes to America*, 76–91, from which this paragraph is extracted, and Edward J. Larson, *Summer for the Gods: The Scopes Trial and America's Continuing Debate over Science and Religion* (New York: Basic Books, 1997). On reaction to the trial in Japan, see G. Clinton Godart, *Darwin, Dharma, and the Divine: Evolutionary Theory and Religion in Modern Japan* (Honolulu: University of Hawai'i Press, 2017), 1–2, 187.

27. Dean Rapp, " 'Such Things Can Only Happen in America': British Press Response to the Scopes Trial," *American Journalism* 7 (1990), 148–63, quote on 149;

Saturday Review, July 25, 1925, 87; undated extract from the *Manchester Guardian*, quoted in "Tennessee's Big Monkey Trial Stirring Europe to Sarcasm at Old-fashioned Ideas of US," *Brandon Daily Sun*, July 13, 1925, 1.

28. "Evolution and Intellectual Freedom," supplement to *Nature* 116 (July 11, 1925): 69–83; J. McKeen Cattell, "Science and Intellectual Freedom," *Nature* 116 (September 5, 1925), 358. See also the coverage of this supplement in "Europe Is Amazed by the Scopes Case," *New York Times*, July 11, 1925, 1–2.

29. "Europe Is Amazed by the Scopes Case," *New York Times*, July 11, 1925, 1, quoting Lloyd George, "Really to Understand Evolution—Read the *Outline of History*," advertisement in *New York Times Book Review*, July 19, 1925, 24; "Shaw and Belloc Debate 'What Is Coming,'" *New York Times*, June 28, 1925, 21; "Shaw Calls Ideas of Bryan 'Infantilism'; without Evolution, He Says, There Is No Hope," *New York Times*, June 10, 1925, 1, which reports that Shaw used the phrase "the stigmata of blockhead." See also George Bernard Shaw, *The Political Madhouse in America and Nearer Home: A Lecture* (London: Constable, 1933), 5–6, and L. Maren Wood, "The Monkey Trial Myth: Popular Culture Representations of the Scopes Trial," *Canadian Review of American Studies* 32 (2002): 147–64.

30. Arthur Keith, "Darwin's Theory of Man's Descent as It Stands To-day," supplement to *Nature* 120 (September 3, 1927): 14–21; "British 'Daytonism' Astonishes Keith," *New York Times*, September 14, 1927, 6; A. G. I., "Those Ignorant Americans," *Evolution: A Journal of Nature* 1 (December 1927): 8. I am indebted to Henrique Caldeira for bringing this article to my attention.

31. "Exit Mr. Scopes," *Brandon Daily Sun*, July 23, 1925, 7.

32. Doug A. Adams, "The War of the Worlds: The Militant Fundamentalism of Dr. Thomas Todhunter Shields and the Paradox of Modernity," unpublished PhD dissertation, University of Western Ontario, 2015, 318 (ridiculous), 369–70 (WCFA, Bryan, debates). See also Colin R. Godwin, "Ignorant Fundamentalists? Ministerial Education as a Factor in the Fundamentalist/Modernist Controversy in the Baptist Convention of Ontario and Quebec, 1927–1933," *Pacific Journal of Baptist Research* 4 (2008): 5–31, and Sam Reimer, *Evangelicals and the Continental Divide: The Conservative Protestant Subculture in Canada and the United States* (Montreal: McGill-Queen's University Press, 2003), 26–28.

33. "Letter to the Editor of Science from the Principal Scientific Authority of the Fundamentalists," *Science* 63 (1926): 259; Numbers, *Creationists*, 89–91.

34. Numbers, *Creationists*, 92 (first).

35. Numbers, *Creationists*, 162–63.

36. Numbers, *Creationists*, 89, 116.

37. See, for example, the advertisement in the *Kingston Daily Gleaner*, August 4, 1925, 2.

38. "Scopes Guilty, Fined $100, Scores Law; Benediction Ends Trial, Appeal Starts; Darrow Answers Nine Bryan Questions," *New York Times*, July 22, 1925, 1 (Bryan); Larson, *Summer for the Gods*, 140 (twenty-two).

39. One of the early reports provided by the Australian Cable Service was "Man's Origin; American Controversy; Teacher on Trial," (Sydney) *Sun*, July 6, 1925, 1; also published in the (Brisbane) *Daily Standard*, July 7, 1925, 7. On July 9 alone, the following appeared: "Darwin Theory: An American Trial," (Perth) *Daily News*, July 9, 1925, 7;

much the same in "Test Case in United States," (Melbourne) *Argus*, July 9, 1925, 11; "The Darwinism Theory; Its Teaching Prohibited; Test Case in America," (Melbourne) *Age*, July 9, 1925, 10; "Evolution Theory: Tennessee's Ban on Teaching," (Brisbane) *Telegraph*, July 9, 1925, 5; "Teaching Darwinism in Violation of Bible: Unique Test Case," (Lismore) *Northern Star*, July 9, 1925, 5; "Interesting Trial," *Newcastle Morning Herald and Miners' Advocate*, July 9, 1925, 5; "Darwinian Theory: Interesting American Trial," *Brisbane Courier*, July 9, 1925, 15; "Teaching of Evolution; Breach of the Law; Remarkable Test Case," (Hobart) *Mercury*, July 9, 1925, 7 (Australian Press Association); "Darwinism Banned; Teacher Defies Law; Strange American Test," (Launceston) *Examiner*, July 9, 1925, 5; "Darwinian Theory Teaching Is Opposed to Tennessee Law: Teacher to Be Prosecuted," (Broken Hill) *Barner Miner*, July 9, 1925, 1; "The Creation; American Test Case; Letter of the Bible and Evolutionary Teaching," (Wagga Wagga) *Daily Advertiser*, July 9, 1925, 2.

40. "Men and Monkeys: Turmoil in Tennessee," (Melbourne) *Advocate*, July 16, 1925, 19.

41. "Interesting Trial," *Newcastle Morning Herald and Miners' Advocate*, July 9, 1925, 5; "Test Case in United States," (Melbourne) *Argus*, July 9, 1925, 11; "Freak Trial; Opened with Prayer; Bible and Politics; Dayton Entertains U.S.A.," (Sydney) *Sun*, July 11, 1925, 1; "Great Monkey Trial: 'Holy War' Starts in Tennessee Town," (Brisbane) *Daily Standard*, July 11, 1925, 1; "America's New Stunt," *Nambucca and Bellinger News*, July 17, 1925, 2; "A Bizarre Trial: And So Typically American; Spirit of Carnival and Fanaticism," (Murwillumbah) *Tweed Daily*, July 13, 1925, 3.

42. "The Debate at Dayton, Tennessee," *Newcastle Sun*, July 11, 1925, 4; "The Land O'Liberty: Do Anything You Like Except Think—Is the Yewnited States Setting Australia a Lead?" (Perth) *Truth*, July 11, 1925, 12; "Science and Religion in America," (Adelaide) *Register*, July 18, 1925, 8; "Evolution: What It Is—and Is Not," (Brisbane) *Worker*, July 23, 1925, 6. See also "America and Evolution: 'Fifty Years behind the Times,'" *Don Dorrigo Gazette*, July 25, 1925, 2.

43. Raymond A. Dart, "*Australopithecus africanus*: The Man-ape of South Africa," *Nature* 115 (1925): 195–99; "Missing Link Theory; On Trial for Preaching Evolution; Comic Opera Case; Scientists Believe Man's Ancestor Has Been Found," *Newcastle Sun*, July 9, 1925, 1.

44. "Tennessee's Heresy Hunt: The 'Down with Evolution' Comedy," *New Zealand Free Lance*, July 15, 1925, 4; "A Circus of Darkness: Blind Fanaticism," *New Zealand Herald*, July 11, 1925, 1; "Man in the Street," letter to the editor, *New Zealand Herald*, July 14, 1925, 7; *Otago Daily Times*, July 30, 1927. See also "Darwinian Theory Trial," *New Zealand Free Lance*, July 13, 1925, 9, which reported the circus-like atmosphere in Dayton, Tennessee; "Law against Darwinism," *New Zealand Herald*, July 9, 1925, 5; and "The Science War in Tennessee," *New Zealand Herald*, July 29, 1925, 32. This and the following three paragraphs have been extracted from Ronald L. Numbers and John Stenhouse, "Antievolutionism in the Antipodes: From Protesting Evolution to Promoting Creationism in New Zealand," *British Journal for the History of Science* 33 (2000): 335–50; reprinted in Simon Coleman and Leslie Carlin, eds., *The Cultures of Creationism: Anti-Evolutionism in English-Speaking Countries* (Aldershot, England: Ashgate, 2004), 125–44. I am grateful to my friend John Stenhouse for his collaboration on this essay.

45. Education Department (New Zealand), *Syllabus of Instruction for Public Schools* (Wellington, 1928), 41–42, 214–15; W. H. Pettit, *Evolution: Is It Scientific? Shall We Teach It in Our Schools*? (Auckland, [1929]), 6, 20. An appendix to Pettit's pamphlet, 23-27, reprints letters on evolution in the schools that appeared in the *New Zealand Herald* in April 1929. For a historical discussion of the controversy, see J. L. Ewing, *Development of the New Zealand Primary School Curriculum, 1877-1970* (Wellington, 1970), 175–86.

46. A. E. C., *The Science of Anti Evolutionists: A Reply to Dr. W. H. Pettit's Pamphlet* (n.p., n.d), unpaginated. This pamphlet was reprinted from the July 1929 issue of *Truth Seeker*, the official organ of the *New Zealand Association for the Advancement of Rationalism*.

47. P. B. Fraser, "'Evolution in Schools': New Syllabus Challenged," *Otago Daily Times*, February 26, 1929, 7; R. M., "Letter to the Editor," *Otago Daily Times*, February 28, 1929, 3; *Lyttelton Times*, June 25, 1929 (only eight persons dissented from the anti-evolution resolution); P. B. Fraser, *Evolution and Ape Manism in Schools: New Syllabus Challenged* (Dunedin, 1929); "Evolution," *New Zealand Education Gazette*, 8 (1929): 141; W. H. Pettit, "Is Man an Improved Ape? Shall We Broadcast the Theory of Evolution?" *Reaper* 15 (1937–38), 67–68. On Fraser, see A. K. Davidson, "A Protesting Protestant: The Rev. P. B. Fraser and New Zealand Presbyterianism 1892–1940," *Journal of Religious History* 14 (1986): 193–217. For the reaction of one Muslim to the Scopes trial, see Alper Bilgili, "An Ottoman Response to Darwinism: İsmail Fennî on Islam and Evolution," *British Journal for the History of Science* 48 (2015): 565–82.

48. The best account of this activity appears in Edward J. Larson, *Trial and Error: The American Controversy over Creation and Evolution*, 3rd ed. (New York: Oxford University Press, 2003), ch. 6.

49. "Creationism in NZ 'Unlikely,'" *NZ Herald*, July 3, 1986, 14 (Gould), clipping courtesy of Carol Scott; *Wisconsin State Journal*, May 7, 2000, 2A; Richard C. Lewontin, introduction to Laurie R. Godfrey, ed., *Scientists Confront Creationism* (New York: W. W. Norton, 1983), xxv; Edward O. Wilson, quoted in Nicholas Wade, "Long-Ago Rivals Are Dual Impressarios of Darwin's Oeuvre," *New York Times*, October 25, 2005, D2. For a similar opinion, see Michael Ruse, *The Evolution-Creation Struggle* (Cambridge, MA: Harvard University Press, 2005), 5.

50. See Numbers, *Creationists*, ch. 18.

51. David Parker, "Fundamentalism Still Fighting for the Faith," *Interchange* 35 (1984): 3–45; Neville Buch, "Americanizing Queensland Protestantism," unpublished MS thesis (1985); John Knight, "Original Arguments: Queensland's Fundamentalists Push for 'Creation Science' Schools," *Australian Society*, July 1, 1984, 18-20; Tony Sperring, "The Evolution of 'Creation Science,'" *Forum of Education* 45 (1986): 23–35; Carl Wieland and Andrew Snelling, "A Creationist Response," *Forum of Education* 47 (1988): 4–20. See also Tony Sperring, "Propagandism, 'Creation Science' and 'Creation Scientists': A Response," *Forum of Education* 47 (1988): 53–86. This discussion of creationism in Australia is largely extracted from Ronald L. Numbers, "Creationists and Their Critics in Australia: An Autonomous Culture or 'the USA with Kangaroos?,'" *Historical Records of Australian Science* 14 (2002): 1–12; reprinted in Coleman and Carlin, *Cultures of Creationism*, 109–23. Regretfully, I have not had the opportunity to read Neville Buch, "American Influence on Protestantism in Queensland since 1945," unpublished PhD thesis, University of Queensland, 1994.

52. Barry Williams to Jim Lippard, April 10, 1991, courtesy of James J. Lippard (kangaroos); Michael Archer, "Evolution as Science: One Aspect of a Very Large Universe," in *Confronting Creationism: Defending Darwin*, ed. D. R. Sekirk and F. J. Burrows (Kensington: New South Wales University Press, 1987), 14–26, quote on 19 (anachronism); "Prove It and They Might Believe It," *The Australian*, June 11, 1990, 9 (Deep South); Barry Price, *The Creation Science Controversy* (Sydney: Millennium Books, 1990), 3 (import); "C.I.A. Funds?" *Ex Nihilo* 8 (1985): 5 (secret funding); "Queensland Schools to Teach Creation!" *Creation Science Prayer News*, June 1984, 1, quoting Phillip Adams in *The Weekend Australian*, May 1984, 12–13 (nonsense). Regarding the legal differences between the United States and Australia, see R. S. Gustafson, "Creation in Australia—The Legal Question," *Ex Nihilo* 3 (1980): 37–40.

53. John Mackay, memorandum dated May 10, 1977, Creation Science Foundation Papers, Answers in Genesis, Acacia Ridge, Queensland; "The Queensland Connection," *Ex Nihilo* 1 (October 1978): 20; "The Queensland Connection," *Ex Nihilo* 3 (May 1980): 1 (Australia's own); Carl Wieland, Letter to the Editor, *Search* 13 (1982): 12 (distinct). On parallel development in New Zealand, see Numbers and Stenhouse, "Antievolutionism in the Antipodes."

54. Jason R. Wiles, "A Threat to Geoscience Education: Creationist Anti-evolution Activity in Canada," *Geoscience Canada* 33 (2006): 135–40, quote on 135 (border); John Barker, "Creationism in Canada," in Coleman and Carlin, *Cultures of Creationism*, 85–108, quote on 96; Numbers, *Creationists*, 404–5. For a popular account of creationism in Canada, see Marci McDonald, *The Armageddon Factor: The Rise of Christian Nationalism in Canada* (Toronto: Random House Canada, 2010), 175–206. See also Phil Plait, "Creationism, Eh?," Bad Astronomy, http://blogs.discovermagazine.com/bad astronomy/2007/10/18/creationism-eh/#.XcrFv1f7Q2w.

55. "Britons Unconvinced on Evolution," BBC News, January 26, 2006, http://news .bbc.co.uk/2/hi/science/nature/4648598.stm; Duncan Campbell, "Academics Fight Rise of Creationism at Universities," *Guardian*, February 21, 2006, https://www.the guardian.com/world/2006/feb/21/religion.highereducation (biology teacher).

56. Michael Gross, "Red Head: US-Style Creationism Spreads to Europe," *Current Biology* 12 (2002), R265–66; N. M. de S. Cameron, "Editorial," *Biblical Creation* 2 (June 1980): 35 (alien element); David C. C. Watson, "Review of *Cross-Currents: Interactions between Science and Faith*, by Colin A Russell," *Biblical Creation* 8 (Spring 1986): 25–30 (politics, any good thing). On creationism in the UK, see also Numbers, *Creationists*, 355–62, 405–8.

57. Debora MacKenzie, "Unnatural Selection," *New Scientist* 166 (April 22, 2000): 35–39; Martin Enserink, "Is Holland Becoming the Kansas of Europe?" *Science* 308 (2005): 1394; Ulrich Kutschera, "Darwinism and Intelligent Design: The New Anti-evolutionism Spreads in Europe," *NCSE Reports* 23 (September–December 2003), 17–18; Numbers, *Creationists*, 399–431. For an excellent recent assessment of European creationism, see Stefaan Blancke, Hans Henrik Hjermitslev, and Peter C. Kjærgaard, eds., *Creationism in Europe* (Baltimore: Johns Hopkins University Press, 2014).

58. Donovan O. Schaefer, "Blessed, Precious Mistakes: Deconstruction, Evolution, and New Atheism in America," *International Journal for Philosophy of Religion* 76 (2014): 75–94.

CHAPTER FIVE

Putting the Structuralist Theory of Evolution in Its Place

NICOLAAS RUPKE

David Livingstone, in a number of important publications, has shown the significance of "location, place and space" for our understanding of the meaning of Darwin and the Darwinian theory of evolution. In his *Dealing with Darwin*, Livingstone compares and contrasts encounters with Darwinism in various centers of Presbyterian thought—Edinburgh, Belfast, Toronto, Decatur (Georgia), and Princeton (New Jersey)—to show how the particulars of place produced a divergence of meaning derived from the reading of Darwin's texts. To put it differently, and to borrow from his *Putting Science in Its Place*, Livingstone has "put Darwin in his place[s]."[1]

This chapter follows up on Livingstone's work by broadening the scope of the subject matter from "Darwinism" to "evolution theory." The two are by no means coextensive, despite misconceptions to the contrary. We can recognize not only different Darwins—that is, multiple appropriations and interpretations of his theory—but also fundamentally different theories of evolution, one of which is the Darwinian, generically speaking.

Scientific theories of evolution have been, in general terms, of three distinct kinds.[2] The first has been characterized by a morphological approach and is more recently referred to as structuralist.[3] This theory, which began taking shape through the second half of the eighteenth century and in the early nineteenth century, interpreted the history of the nonliving and the living world as a process of progressive self-organization driven by the structural properties of matter under changing environmental conditions.[4] Arguably, the outstanding early representative of this theory was Immanuel Kant, who, in his anonymously printed *Allgemeine Naturgeschichte und Theorie des Himmels* (*General History of Nature and Theory of the Heavens*, 1755), put forward the nebular hypothesis of planetary evolution, complemented by a last chapter on (extraterrestrial) life.[5] The *Allgemeine Naturgeschichte* was a product of Kant's precritical scientific writings and published when he was only thirty years old. Parts of Kant's *Kritik der Urteilskraft* (*Critique of Judgment*, 1790) also proved formative, especially an iconic passage on evolutionary morphology:

> It is praiseworthy by the aid of comparative anatomy to go through the great creation of organised natures, in order to see whether there may not be in it something similar to a system and also in accordance with the principle of production. For otherwise we should have to be content with the mere principle of judgement (which gives no insight into their production) and, discouraged, to give up all claim to *natural insight* in this field. The agreement of so many genera of animals in a certain common schema, which appears to be fundamental not only in the structure of their bones but also in the disposition of their remaining parts,—so that with an admirable simplicity of original outline, a great variety of species has been produced by the shortening of one member and the lengthening of another, the involution of this part and the evolution of that,—allows a ray of hope, however faint, to penetrate into our minds, that here something may be accomplished by the aid of the principle of the Technik (mechanism of nature) (without which there can be no natural science in general). This analogy of forms, which with all their differences seem to have been produced according to a common original type, strengthens our suspicions of an actual relationship between them in their production from a common parent, through the gradual approximation of one animal-genus to another—from those in which the principle of purposes seems to be best authenticated, *i.e.* from man, down to the polype, and again from this down to mosses and lichens, and finally to the lowest stage of nature noticeable by us, viz. to crude matter. And so the whole Technic [mechanical workings; structural constitution] of nature, which is so incomprehensible to us in organised beings that we believe ourselves compelled to think a different principle for it, seems to be derived from matter and its powers according to mechanical laws (like those by which it works in the formation of crystals).[6]

Several other late eighteenth- and early nineteenth-century proponents of the structuralist approach are cited later in the chapter.[7]

The second of the three theories, famously put forward by Charles Darwin in his *On the Origin of Species* (1859), attributed life's diversity to the external, ecological factor of competition between members of species. The third theory, younger still, was formulated as part of the Marxist critique of Darwinism—a critique that initially emphasized cooperation over competition and later stressed the heritability of acquired characteristics (arguments that I develop later in the chapter).

The least known today of the three theories is the structuralist, the first here identified. I argue that this is due less to scientific reasons than it is to geographically circumscribed political ones. Through the nineteenth century, evolution theory became increasingly entangled with the politics of nationalism. This process culminated in, and was part of, World War I, the concurrent Russian Revolution, and the dozen years of Nazi domination in Germany. I suggest that the renown of the structuralist approach was affected negatively by a nationalistic identification of evolutionary structuralism with the German cause.[8]

This chapter examines the placed nature of structuralist views of evolution by looking at such views at the scale of the nation, at individual texts, and at the political circumstances of particular authors. If, as I suggest, this is a story of a distinct sort of evolutionary theory over time in which developments in the nineteenth century had enduring significance in the first half of the century following, it is also a story of specific institutional sites and political proclivities.

The chapter is organized in four sections. The first explores the ways in which evolutionary theory in general, and Darwin's ideas on local adaptation and struggle in particular, became politicized—that is, were read and interpreted in particular national contexts. Darwinism in this sense was a peculiarly British phenomenon. The second examines how structuralist evolutionary theory was primarily German in origin and how this theory was regarded as part of German science as it came to be judged, morally and cognitively, in the context of conflict with Germany during World War I. The third section considers what we might think of as the longer-run and political appropriation in the first half of the twentieth century of nineteenth-century evolutionary thought—namely, the uptake by Nazi thinkers in Germany of these ideas and the "competition" between different evolutionary theories in order to serve a particular race-based biology and politics. The appropriation by the Nazis of the structuralist theory of organic evolution could not—and did not—survive World War II. As I show in the fourth section, the "denazification" of this theory meant, in part, a return to nineteenth-century debates in order to serve the new Germanies, West and East, of the mid-twentieth century.

The Early Nationalization of Evolution Theory

Through the second half of the nineteenth and first half of the twentieth century, the conflict surrounding theories of evolution increasingly took place in the sociopolitical arena, rather than that of science and theology. One indication and expression of this was that each of the three main approaches to evolution became nationalized. The first to be explicitly and persistently characterized in national terms was Darwin's theory. Its typecasting in these terms was the work of a growing group of Marxist intellectuals who would become engaged in the Russian Revolution and in Soviet-era affairs. Soon after the appearance of *On the Origin of Species*, Karl Marx and Friedrich Engels noted with disapproval that Darwin's model of survival of favored races in the struggle for life was based on what they considered to be the principles of bourgeois, capitalist economics, as (in)famously formulated by English clergyman and political economist Thomas Malthus in his *An Essay on the Principle of Population* (six editions between 1798 and 1826).[9] On the one hand, Marx adopted Darwin's theory as the scientific foundation of his dialectical materialism. On the other, he disliked its English Malthusianism, initiating a "Darwin without Malthus" trend popular among socialist thinkers ever since. As early as 1862, Marx confided to Engels: "It is curious how Darwin recognizes among beasts and plants his English society, with its division of labor, competition, opening up of new markets, 'discoveries' and Malthusian 'struggle for existence.' It is Hobbes' bellum omnium contra omnes, war of all against all, and is reminiscent of Hegel in the *Phenomenology*, where bourgeois society represents a 'spiritual animal kingdom,' while with Darwin the animal kingdom reflects bourgeois society."[10] Engels held similar views. In his unfinished *Dialektik der Natur*, Engels contended that the "struggle for life" should be seen as counterbalanced by cooperation, in nature as well as in human society: "The entire Darwinian theory of the struggle for life is simply the transfer from society into living nature of Hobbes' notion of bellum omnium contra omnes and of bourgeois competitive economics as well as the Malthusian theory of population."[11]

In the lead-up to the Russian Revolution, a number of Marxist Russians with biological expertise wrote non-Darwinian, alternative evolution treatises in which socialist cooperation was favored over capitalist competition.

Particularly well known in this respect was the work of the Russian revolutionary and geographer Pyotr A. Kropotkin in his *Mutual Aid: A Factor of Evolution* (1902; originally written during the 1890s as a series of articles for *The Nineteenth Century*), which remained in circulation through much of the twentieth century. It was "but one expression of a broad current in Russian evolutionary thought that pre-dated, indeed encouraged, his work on the subject and was by no means confined to leftist thinkers."[12]

Among later leftists were Boris M. Kozo-Polyansky with his *Symbiogenesis: A New Principle of Evolution* (Russian original 1924; English translation, edited by Lynn Margulis, 2010), while amid those accused of right-wing sympathies was yet another Russian critic of Malthusian Darwinism, Lev Berg, whose *Nomogenesis or Evolution Determined by Law* (Russian original 1922; English translation 1924) carried an introduction by Scottish biologist and polymath D'Arcy Thompson (a 1969 edition was introduced by Theodosius Dobshansky).[13] Unlike Kropotkin, however, Berg not only expressed skepticism about the evolutionary efficacy of the Darwinian struggle for life; he mobilized a variety of morphological/structuralist arguments in opposition to Darwin and, in particular, made much of the phenomenon of convergence.

This seemed to amount to siding with German idealism and thus to undermining Marxist thought. Berg's daughter, Raissa L. Berg, herself an accomplished geneticist during the Soviet era, explained how her father's opposition to Darwinism got him into political trouble: "Father refused to recognize the Darwinian principle of natural selection and of the struggle for existence as the reason underlying the progressive evolution of the organic world." Berg was censured for his idealism during these early years of the Soviet Union, as was, in retrospect, Jean-Baptiste de Lamarck. "Darwin, on the other hand, had been made a Marxist saint and to criticize him was to encroach on Marxism's holy of holies. You weren't even to mention the mutual help that rules in nature. Idealism: the weapon of the enemy, the subversion of Marxism, the camouflaged desire to discredit the idea of class struggle—that's what discussions of mutual help were."[14]

The Marxist pronouncements found a sympathetic ear with a number of British scientists, not least several professors and students at the University of Cambridge. These "Cambridge mavericks" included J. B. S. Haldane, J. D. Bernal, and Joseph Needham, and, subsequently, historian of science Robert M. Young, who, in his lectures and their published version, *Darwin's Metaphor* (1985), made the formative significance of Malthusian political economy for Darwinian theory part of mainstream opinion. As John C. Greene remarked:

> It is a curious fact that all, or nearly all, of the men who propounded some idea of natural selection in the first half of the nineteenth century were British. Given the international character of science, it seems strange that nature should divulge one of her profoundest secrets only to inhabitants of Great Britain. Yet she did. The fact seems explicable only by assuming that British political economy, based on the idea of the survival of the fittest in the marketplace, and the British competitive ethos generally predisposed Britons to think in terms of competitive struggle in theorizing about plants and animals as well as man.[15]

Evidence of the nationalization of evolutionary theory at a specific geographical scale—that of the nation—should not blind us to the many differences within nations or to the political and intellectual purposes served by adherence to one theory, or its implications, rather than to another, as Livingstone has shown. Nevertheless, it is helpful to address the national expression of evolutionary theory in order to expose where and how more particularly, and quite why, different theories had the national-political resonance they did.

Germanification of Structuralist Evolution Theory

The structuralist evolution theory was primarily German in origin and its morphological tradition indebted to idealist thought. As mentioned previously, Kant, who taught at Königsberg, produced two seminal treatises in this context—namely, the *Allgemeine Naturgeschichte* and the *Kritik der Urteilskraft*—both of which influenced evolutionary thought in Germany through much of the nineteenth century. Largely in agreement with Kant was Johann Friedrich Blumenbach, the internationally celebrated professor of natural history at the University of Göttingen who in his *Über den Bildungstrieb* (*On Generation*, 1789; second edition 1791) and in arguing for a natural origin of life and species in 1791 put forward the idea of a "Bildungstrieb," or *nisus formativus*, understood by many as a Newtonian force that expressed itself in the generation and regeneration of organic form.[16] Similar views to those of both Blumenbach and Kant were expressed in Alexander von Humboldt's *Cosmos* (first two volumes 1847, 1849), which promoted a naturalistic progressivism that holistically bracketed the sidereal with the terrestrial.

The Blumenbach-centered "Göttingen School" exerted considerable influence on German Romantic biology, synergizing with the work of Johann

Friedrich Meckel at Halle, Karl Ernst von Baer at Königsberg, or Carl Friedrich Kielmeyer at Tübingen. Bringing embryology to bear on taxonomy and paleontology, these men contributed to the notion of recapitulation—a notion that found its late nineteenth-century expression in the biogenetic law (ontogeny recapitulates phylogeny—akin to today's evo-devo) as formulated by Ernst Haeckel in his *Generelle Morphologie der Organismen* (*General Morphology of Organisms*, 1866).[17] Haeckel taught at Jena, another stronghold of the structuralist tradition, where already in the late eighteenth and early nineteenth centuries, Weimar's Johann Wolfgang von Goethe had made foundational contributions to comparative morphology and in his *Metamorphose der Pflanzen* (*Metamorphosis of Plants*, 1790) had formulated the concept of a dynamic archetype. Other ranking names in the structuralist tradition range from Johannes Müller in Berlin to August Weismann in Freiburg im Breisgau.

Moreover, several of the non-German advocates of this approach, including Étienne Geoffroy Saint-Hilaire in France and Richard Owen in England, were known Germanophiles who aligned themselves with the tenets of German Romanticism and idealist philosophy. Recently, the argument has been advanced, most eloquently by Robert J. Richards, that Darwin, too, was substantively indebted to German Romantic biology.[18] Additional names in this enumeration—names that, if not cited here, might be missed by informed readers—are the Frenchman Jean-Baptiste de Lamarck and the German Lorenz Oken. Both had much in common with the structuralists, but they additionally indulged in nature-philosophical speculations that went well beyond the restraints of the Kant-Humboldtian epistemology of methodological naturalism.

Of course, if Richards is right that Darwin, in formulating the theory of descent with modification by natural selection, was significantly influenced by German thought, then this would undermine the distinction between a German structuralist evolutionary tradition and a British functionalist one. A major argument used by Richards is that Darwin's notion of evolution owed much to Humboldt's holistic view of the world as expressed in the *Personal Narrative of Travels* (7 volumes, 1819–29). Whereas Humboldt's holism was Kantian and structuralist, Darwin's holism—his belief in the evolutionary relatedness of all living beings—was of a different kind. Humboldt's *Personal Narrative* as well as his *Cosmos* outlined a sketch of a harmonious integrated universe that intimately bound together the physical with the biological—the stellar, the planetary, the terrestrial with the living world, including humans and the very human culture that was producing the late eighteenth/early nineteenth theory of evolution itself.

Nothing like that was to be found in *On the Origin of Species* (1859), which, by stark contrast, hammered a wedge between the nonliving and the living world, in the sense that Darwin ignored the Kant-Humboldtian theory of progressive development in the physical, nonliving sphere. Most strikingly, Darwin belittled or altogether ignored the fundamental question of the origin(s) of life, because his evolutionary mechanism of natural selection can only operate once living systems are in existence. To Humboldt, who died the year Darwin's magnum opus was published, the notion of natural selection neither could nor would have been acceptable as anything more than an addition to universal progressive development. To him, the evolutionary complexification of the world was a lawlike response of matter and its properties to environmental conditions, beautifully exemplified by the regularities of the zonal distribution of plants across the globe.[19]

The fact that structuralist evolution theory emanated first and foremost from German universities turned into a public issue and an item of political contention when it became mixed up in the war of words that accompanied World War I. The nationalistic hostility and polarization between the Germans and their adversaries carried over into the academic world. A so-called mobilization of intellect took place in Germany, France, and Great Britain. Manifestations included the German *Aufrufe*, public declarations or manifestos by academics who avowed support for the Reich's military and proclaimed Germany's moral, scholarly, and scientific superiority.[20]

In the wake of the Franco-Prussian War (1870), the Germans had already credited their victory to "die Deutsche Wissenschaft" (German science). Such scientific superiority was laid claim to once more during World War I in the form of impressive lists of signatories attached to various scientific and political declarations. Most (in)famous was "An die Kulturwelt! Ein Aufruf" (1914), colloquially known as "The Manifesto of the 93," as it was signed by no fewer than ninety-three stars in the German academic firmament, including Max Born, David Hilbert, Max Planck, and biologists such as Haeckel. Another manifesto, the "Erklärung der Hochschullehrer des Deutschen Reiches" ("Declaration of the University Professors of the German Reich") featured no fewer than 3,082 professors—a few were not full professors—of Germany's fifty-three institutions of higher learning. Among the rare abstentions was Albert Einstein, who, together with only two others, signed the antiwar "Aufruf an die Europäer" ("Appeal to the Europeans," 1914), composed by left-leaning physiologist Georg Friedrich Nicolai. The manifesto remained unpublished until 1917, when it was included in Nicolai's *Biologie des Krieges* (*Biology of War*)—an indictment of warfare and a critique of Darwin's Malthusian thought.[21]

Less well known is that the British and the French counterpunched with considerable rhetorical ferocity. *Les Allemands et la Science* (*The Germans and Science*, 1916) was a collection of no fewer than twenty-eight chapter-length denunciations of German humanities and sciences, written by a distinguished group of university professors and initiated at the *Académie des sciences* and those other academies that composed the Institut de France. This collaborative publication added up to what probably was the single largest, most comprehensive, most emotional, and most abusive attack ever on the academic learning of any one nation. The Neanderthal expert Marcellin Boule denigrated German paleontology; Stanislas Meunier, famous for his experimental approach, did the same for geology. British scientist William Ramsay, who was awarded the Nobel Prize in Physical Chemistry in 1904, deprecated Germany's "mediocre" contributions to scientific discovery; physicist and historian of science Pierre Duhem distilled a series of his university lectures to form a chapter on the inferiority of Germanic/Teutonic science. Others singled out Ernst Haeckel for attack while praising Darwin, even though Haeckel had often placed Darwin at the top of his list of scientific "heroes." Louis-Félix Henneguy, at the Collège de France, denounced not only Haeckel but also the venerable August Weismann and the latter's notion of chromosomes as carriers of hereditary features. Henneguy referred to the "demented megalomania" of the Germans and concluded that their biology—indeed, their science in general—was untrustworthy: "Nous devons considérer comme suspectes les publications germaniques, comme tout objet *made in Germany*"[22] ("We have to distrust German publications, like everything *made in Germany*"). Duhem's article drew upon his fuller-length pamphlet to the same anti-German intent, *La Science Allemande* (*German Science*, 1915), which remained in circulation until well after World War II. As a devout Catholic, he was not a follower of Darwin. Combining his faith with patriotism, he instead expressed warm approval of fellow Catholics and countrymen Henri Fabre and Louis Pasteur while painting Ernst Haeckel in a negative light.[23]

In the wake of the Great War, multiple instances took place of German science and its representatives being banished from international meetings. The German language, which by the turn of the century had nearly attained the status of lingua franca of academic communication, was jettisoned in favor of English, and on the coattails of the avoidance of German in international scholarly communication rode a new dislike of, and in places open disdain for, "die deutsche Wissenschaft."[24] Haeckel died in 1919, less than a year after the end of what already in 1914, all too prophetically, he had termed "the First World War." The structuralist approach now carried the less fashionable label "Germanic" or "Teutonic," and its international popularity declined along with German cultural and political sway.

Nazi Appropriation

Worse was to follow during the period of the Third Reich and World War II in that Haeckel's German heroes of evolutionary thought—Goethe and Kant—were appropriated by the Nazis. The complex relationship between Darwinism and the Nazis was apparent at a number of levels. At the popular level, at which Adolf Hitler and much of the National Socialist leadership operated, Darwinism first and foremost meant social Darwinism and eugenics.[25] The name "Darwin," inasmuch as it denoted "the preservation of favored races in the struggle for life," or even Herbert Spencer's more catchy "survival of the fittest," were unproblematic terms.

During the Third Reich, Ernst Haeckel's "Erbe" (heritage) was reinterpreted by leading Nazi scientists, especially for racist biological purposes.[26] The Nazis particularly approved of the way in which Darwinism, in its New Synthesis guise, incorporated some of the anti-Marxist genetics of Germanic biology. In the official science textbook for soldiers who studied for their "Reifeprüfung" (high school diploma), commissioned by the "Oberkommando der Wehrmacht," Darwinism was put forward as the basis for a race-based biology:

> Apart from small weaknesses and problems that need not be mentioned here, Darwin's theory stands its ground against the modern scientific critique, especially from the side of genetics. His fundamental notions, such as hereditary variations, selective breeding, struggle for life, adaptation, natural selection, are scientifically unobjectionable and clear. They are capable of explaining from a single principle a wealth of facts and of bringing them under a common denominator. . . . Next to genetics, Darwinism provides us ideologically and politically with the foundation for the biological and race hygienic measures in today's state. . . . In this way racial hygiene wants to prepare the soil for natural selection also among humans, after false humanitarianism and a failure to recognize biological connections and necessities had made this almost ineffectual.[27]

At a different level, however, among the more erudite Nazi ideologues and biological scientists, Darwinian evolution theory became typecast as philosophically unsound and unworthy of being a scientific *Leitkultur* (dominant culture). Even by the start of the twentieth century, Houston Stewart Chamberlain, the British-born naturalized German whom Hitler lauded as "the prophet of the Third Reich," had criticized Darwin's theory for being rooted in Enlightenment materialism and humanism, a "fault" equally attributed to Bolshevist socialism. As the pan-Germanic and anti-Semitic Chamberlain argued in his influential *Die Grundlagen des Neunzehnten Jahrhunderts* (*Foundations of the Nineteenth Century*, 1899), Darwin was a product of the "shallow" rationalism of the French Enlightenment. Admittedly, Chamberlain did not think too highly of Haeckel, either, yet he greatly admired Goethe and Kant. The German approach to the origin of species should be seen as a product of Germanic "Kulturgut" (cultural heritage), which could trace its roots to Weimar classicism and the idealism of the Romantic period.

> In Herder, Kant and Goethe we meet with the idea of evolution in characteristic colouring; it is the revolt of great minds against dogma: in the case of the first, because he, following the course of Teutonic philosophy, endeavoured to find in the development of the idea "nature" an entity embracing man; in the case of the second, because he as metaphysician and moralist could not bear to lose the conception of perfectibility, while the third, with the eye of the poet, discovered on all sides phenomena which seemed to him to point to a primary relationship between all living organisms, and feared lest his discovery should evaporate into abstract nothingness if this relationship were not viewed as resting upon direct descent.[28]

By contrast, "Darwin's system was 'manifestly unsound' and even philosophically inadequate," based as it was on "practicability" and "specialization." In his equally voluminous *Immanuel Kant* (1905), Chamberlain criticized the Darwinian overemphasis on function, as opposed to Lebensgestalt (organic form): "Darwin himself contributes a lot to this confusion by fundamentally avoiding every philosophical discussion of the concept that lies at the root of all of his thoughts. For example, if thinking had not become so unfashionable, every human being would have to be highly surprised not to find a single discussion of the concept of "species" in a book that carries the title *On the Origin of Species*."[29]

Other National Socialist scholars and scientists elaborated upon this Nazification of the structuralist tradition. One biography of Alexander von Humboldt accentuated Humboldt's debt to Goethe's idealism, and Goethe and Humboldt were together presented as precursors of National Socialism.[30] Several structural evolutionists, such as the Austrian cofounder of paleobiology Othenio Abel, were outspoken anti-Semites and active participants in Nazi politics. The botanist Wilhelm Troll, best known for his inflorescence studies, combined in his botany idealist morphology in the tradition of Goethe with political proximity to Hitler.[31]

Leading Third Reich scientists concurred. Among them was the politically active paleontologist Karl Theodor Beurlen. Working under the direction of the multifunctionary "Reichsstudentenführer" Gustav Adolf Scheel, Beurlen was a key figure in the Nazification of scientific textbooks, and through his work in this respect, structuralist evolution theory acquired the stamp of Party approval. As a geologist, he was more interested in Charles Lyell and in his influence on Darwin than in Darwin himself. While he acknowledged the scientific quality of Lyell's *Principles of Geology* and its innovative actualistic uniformitarianism, Beurlen argued for the historicist linkage of *Erdgeschichte* (history of the earth) with *Lebensgeschichte* (history of life)—that is, for the nonrecurrent uniqueness of each successive phase in the evolution of Earth and life.[32]

The person who, so to say, "liberated" geology from the rationalistic cosmogony of the Enlightenment was Karl Ernst Adolf von Hoff. In his *Geschichte der durch Überlieferung nachgewiesenen natürlichen Veränderungen der Erdoberfläche* (*History of Natural Changes of the Surface of the Earth as Known from Human Records*, 5 volumes, 1822–41), von Hoff had, like Lyell, put forward the principle of actualism but, contra Lyell, without the assertion of uniformitarianism. This was a forebear to Chamberlain's critique: Lyell and Darwin were products of Enlightenment rationalism and its fundamentally ahistorical Huttonian approach to geology and to paleontology, while German biology was, by contrast, steeped in the historicism of Johann Gottfried Herder.

Yet, as pointed out by Ute Deichmann in her work on the biological sciences in the Nazi regime, no fully fledged "German biology," similar to "German physics," was ever formulated or institutionalized despite efforts by Tübingen botanist Ernst Lehmann.[33] Considerably more powerful than Lehmann was the zoologist, geneticist, and anthropologist Gerhard Heberer. An early recruit to National Socialist organizations, Heberer joined the National Socialist German Workers' Party in 1937 and, later, the SS (Schutzstaffel, the paramilitary wing of the Nazi Party), rising in 1942 to the rank of Haupsturmführer. With the personal support of Heinrich Himmler, he was given a position in the Rasse- und Siedlungshauptamt (Race and Settlement Main Office) of the SS for the purpose of checking the genealogy of its members and the racial purity of their marriages. Heberer also became a member of Himmler's influential research foundation Deutsches Ahnenerbe (German Ancestral Heritage), which furthered Aryan race ideology and the Germanization of cultural acquisitions. From 1938 until 1945, he was professor of biology and anthropogenesis at the University of Jena, where, together with the notoriously racist and eugenicist rector of the university Karl Astel, as well as others, Heberer formed the so-called Rassen-Quadriga and contributed to Astel's effort of turning Jena into the SS-Muster-Universität (SS Model University).[34]

Heberer contended that evolutionary biology was fundamental to one's worldview in general and to "Staatsbiologie" in particular. His major contribution to the subject was a collected volume titled *Die Evolution der Organismen* (1943), which brought together a team of nineteen leading biologists and paleontologists that included Konrad Lorenz, Viktor Julius Franz, Nikolay Timofeev-Ressovsky, and Walter Zimmermann, each of whom was a Nazi collaborator and a supporter of Aryan race biology. Like Astel and Brücher, they appropriated Haeckel's heritage and wrote Goethe's name on their banner. What we may think of as the Anglo-American New Synthesis was given a warm reception, especially by Timofeev-Ressovsky. This New Synthesis combined the struggle for life with the adoption, albeit partial, of genetics, which suited National Socialist racist thought: "These insights oblige us, make it our duty, to ensure by means of intentional selection a people of racial quality, genetically healthy, and culturally creative" ("Diese Erkenntnisse verpflichten, stellen uns die Aufgabe, in bewußter Zuchtwahl für ein rassisch tüchtiges, erbgesundes, kulturschöpferisches Volk der Zukunft zu sorgen").[35]

Above all, the volume celebrated the dominant arrival in evolution theory of the genetics of Gregor Mendel, Hugo de Vries, Carl Correns, and others. As Heberer proclaimed, "The genetic conception of life has become a fixed foundation stone of the scientific-biological worldview" ("Die genetische Auffassung des Lebens ist ein fester Grundstein des naturwissenschaftlich-biologischen Weltbildes geworden").[36] He repeatedly emphasized that evolutionary biology had become genetics, the new and firm basis for "Abstammung," for descent, for phylogeny. This was a volume about phylogenetics, written during the victorious early years of the war: "The work has been written in the middle of the European struggle for freedom. It is however not merely a book of the home country, because several contributors have written their chapters as soldiers and did not forget their work even during the action at the front! . . . Thus the book is also a gift of the fighting front!" ("Das Werk ist inmitten des europäischen Freiheitskampfes geschrieben worden. Es ist aber nicht nur rein Buch der Heimat; denn mehrere Mitarbeiter haben ihre Beiträge als Soldaten verfaßt und selbst während des Fronteinsatzes die Arbeit nicht vergessen! . . . So ist das Buch zugleich auch eine Gabe der kampfenden Front!")[37]

Denazification

Following Germany's defeat in World War II, to put forward a structuralist theory of organic evolution, one besmirched by having the fingerprints of the Nazis, their collaborators, and sympathizers all over it, could prove inopportune to put it mildly. Both in West and East Germany, many earth and life scientists kept their distance from the structuralist views of Chamberlain, Beurlen, Heberer, and their associates. Beurlen, after his release from detention, left Europe for South America, together with many other unrepentant Nazis. Heberer, by contrast, after completing a two-year jail sentence, more demonstratively joined those who either took the side of the Anglo-American Darwinians or asserted that they had been on the side of the New Synthesis all along. Freshly "entnazifiziert" (declassified as a Nazi), he declared himself affiliated with neo-Darwinism in a booklet entitled *Was heißt heute Darwinismus?* (*The Meaning of Darwinism Today*, 1949). Heberer proved immovable from this latter course of intellectual action and when, at the time of the centenary of the *On the Origin of Species* in 1959 Western biology celebrated Darwin and his ascension to hegemony, Heberer payed homage by "re-educating" West Germans through his *Charles Darwin: Sein Leben und sein Werk* (*Charles Darwin: His Life and Work*, 1959). Being a Darwinian in post–World War II Germany and in neighboring countries whose scientists had collaborated with the Nazis became something of a "Persilschein"—that is, a certificate that helped cleanse one of Nazi "dirt and smudges," or, no less simply, a means of distancing oneself from the recent political past and joining the cultural traditions of the victorious Allies. British Darwinians, in turn, were now able to denigrate or ignore structuralist thought and to do so on the basis of its German connotations.[38]

A cautious attempt was made by some members of the older generation of scholars and scientists to salvage parts of idealist culture and the structuralist tradition. Wolf von Engelhardt, mineralogist and Goethe scholar from Tübingen University, made an elegant effort to perpetuate and rehabilitate the structuralist tradition. His success was limited, however, as the journal in which he and others published, *Studium generale*, did not survive long, even though its editors increasingly permitted the inclusion of contributions in the English language. Another Tübingen figure, Otto Schindewolf, the formidable champion of "a saltational, internally driven evolutionism in the Continental formalist tradition"—who, unlike Abel, had not blotted his political copy book during the Third Reich—was unable to keep the younger generation of paleontologists and historians of biology, including Tübingen's Wolf-Ernst Reif and Thomas Junker, from turning to "the longstanding English preference for functionalist theories based on continuous adaptation to a changing external environment."[39]

The new "Darwin," "the Darwin of the twentieth century," was Harvard's Ernst Mayr, whose German background was integral to his efficacy in helping Darwinize West German evolutionary biology post–World War II. Even Adolf ("Dolf," for obvious reasons) Seilacher's inventive *Konstruktionsmorphologie* (construction morphology) was met with dismissive puzzlement by Princeton's German-American paleontologist Alfred ("Al") Fischer.[40] The structuralist tradition faded from mainstream scientific discourse in both East and West. To the extent that a memory of it was alive at all, it fell to the historians of science. Adolf Meyer-Abich's *Biologie der Goethezeit* (*Biology during the Time of Goethe*, 1949) was an instance of this, but it did not trigger much of a following, as his credibility suffered from having been one of the signatories to the *Bekenntnis der deutchen Professoren zu Adolf Hitler* (*Confession of Faith in Adolf Hitler by German Professors*, 1933).

In East Germany, too, a stated preference for Darwin functioned as a means of distancing oneself from the National Socialist past. Yet a different version of Darwinian theory was used. Where the West Germans were keen to put in place *Wiedergutmachung* (reparation) measures that matched beliefs and values in the Western, Anglo-American world of free market capitalism, the East Germans had to accommodate the Soviet victors and their Marxist ideology.[41] This meant that the structuralist element of genetics, as present in the neo-Darwinian New Synthesis and sanctioned by the Nazis, was rejected in the East. Instead, a so-called *schöpferischer Darwinismus* (creative Darwinism) became the catchword in the German Democratic Republic. This involved the rehabilitation of Lamarck, albeit stripped of his idealist leanings, and a merger with Darwinism by means of emphasizing the importance of environmental factors.

Conclusion

How are we to understand these geographies of evolution theory? The overriding feature of my story is a spatialization phenomenon in which boundaries were drawn around different solutions to the scientific problem of origins by nationalism and war. The parameters that determined the provinces of distribution of the different evolution theories were identical to those that determined the geographical borderlines of the warring nations. This insight helps recover the remarkable extent to which the theories, and in particular the structuralist approach, were formatively embedded in the matrix of national culture and in particular politics. The main reason for the mid-twentieth-century decline of the structuralist theory of evolution was its distinct sociopolitical province of distribution. In other words, the geographical coordinates, national location, and political space proved of consequence to its apparent scientific validity or lack of it. The demise of the structuralist theory of evolution is linked tightly to the theory's German location and to Germany's defeat in World War I and World War II.

If this thus demonstrates the significance of place and space in the historiography of evolutionary thought—of thinking geographically about the production, mobility, and reception of different theories within the biological sciences—it is also to recognize how ideas that had particular expression in the nineteenth century drew upon earlier notions and could have later and lasting repercussions. What is incumbent upon historians—and geographers—of science is not to separate the chronology of ideas from their geography but, rather, to make clear how, where, why, and by whom ideas were shaped and received, and with what result.[42]

Notes

1. David N. Livingstone, *Dealing with Darwin: Place, Politics, and Rhetoric in Religious Engagements with Evolution* (Baltimore: Johns Hopkins University Press, 2014); *Putting Science in Its Place: Geographies of Scientific Knowledge* (Chicago: University of Chicago Press, 2003).

2. I deal with broad categories of evolutionary theory that have found large followings, and not with idiosyncratic, and for the most part individualistic, interpretations of evolution. For a variety of alternative evolution theories, see Georgy Levit, Kay Meister, and Uwe Hossfeld, "Alternative Evolutionary Theories: A Historical Survey," *Journal of Bioeconomics* 10 (2008): 71–96.

3. For more on the terms "structuralist"/"structuralism" in evolutionary biology, see, among others, G. Webster and B. C. Goodwin, "The Origin of Species: A Structuralist Approach," *Journal of Social and Biological Structures* 5 (1982): 15–47. It contains much of value, but the choice by them of Hans Driesch as a typical structuralist is, in my opinion, infelicitous.

4. At the time, structuralist evolution theory was not known by this name nor was it known as such during much of the nineteenth century. The absence of the name need not subtract, however, from its conceptual distinctiveness. Indeed, the term "evolution," in the sense of "the natural origin of species," did not become current until the second half of the nineteenth century, and before that had the creationist meaning of "preformation," even in the first edition of Darwin's *On the Origin of Species*. The more commonly used contemporaneous terms for what we now refer to as "structuralist evolution" was "law-like (progressive) development" or "nomogenesis," the latter term being coined by Richard Owen; see Nicolaas Rupke, *Richard Owen: Biology without Darwin* (Chicago: University of Chicago Press, 2009), 173, 177. Further, "formalist" or "morphological" have been in circulation as synonyms with "structuralist," which itself only gained currency after 1945.

5. Nearly the entire print run of this ill-fated treatise was lost when the publisher went bankrupt at the time of publication. A later abstract presented a truncated summary of the contents, leaving out the essential last chapter that dealt with the inevitability of life, and life on other planets. See Andreas Losch, "Kant's Wager: Kant's Strong Belief in Extra-terrestrial Life, the History of This Question and Its Challenge for Theology Today," *International Journal of Astrobiology* 15 (2016): 261–70.

6. *Kant's Critique of Judgment*, trans. J. H. Bernard, 2nd ed. (London: Macmillan, 1914), para. 80, 337–38; emphasis original.

7. Popular and semipopular renditions included the anonymously published *Vestiges of the Natural History of Creation* (1844) by Robert Chambers and Hermann Burmeister's *Geschichte der Schöpfung* (*History of Creation*, 1845).

8. The existence and importance of national research traditions in evolutionary biology, especially with respect to Germany's contributions, are discussed in detail by Georgy Levit and Uwe Hossfeld in "Major Research Traditions in Twentieth-Century Evolutionary Biology: The Relations of Germany's Darwinism with Them," in *The Darwinian Tradition in Context: Research Programs in Evolutionary Biology*, ed. R. G. Delisle (Dordrecht: Springer, 2007), 169–93.

9. On Marx and Engels on Malthus, see Robert J. Mayhew, *Malthus: The Life and Legacies of an Untimely Prophet* (Cambridge, MA: Harvard University Press, 2014), 141–42.

10. Karl Marx and Friedrich Engels, *Briefwechsel/Correspondence*, vol. 3, 1861–1867 (Berlin: Dietz, 1950), 94–95 (author's translation).

11. Friedrich Engels, *Dialektik der Natur* (Berlin: Dietz, 1952), 328 (author's translation).

12. Daniel P. Todes, *Darwin without Malthus: The Struggle for Existence in Russian Evolutionary Thought* (New York: Oxford University Press, 1989), 104.

13. On the nomogenetic theory, see Georgy Levit and Uwe Hossfeld, "Die nomogenese: Evolutionstheorie jenseits von Darwinismus und Lamarckismus," *Verhandlungen zur Geschichte und Theorie der Biologie* 11 (2005): 367–88. See also Armin Geus and

Ekkehard Höxtermann, eds., *Evolution durch Kooperation und Integration: Zur Entstehung der Endosymbiosetheorie in der Zellbiologie* (Marburg an der Lahn: Basilisken, 2007). On D'Arcy Thompson's own work in this respect, see his influential *On Growth and Form* (Cambridge: Cambridge University Press, 1917). Thompson did not openly reject Darwin's ideas of natural selection but, rather, regarded it as secondary to questions of origin in biological forms. He was thus a proponent of structuralism in the ways I have defined it here. On D'Arcy Thompson's work, see Alan Werritty, "D'Arcy Thompson's " 'On Growth and Form' and the Re-discovery of Geometry within the Geographic Tradition," *Scottish Geographical Journal* 126 (2010): 231–57.

14. Raissa L. Berg, *Acquired Traits: Memoirs of a Geneticist from the Soviet Union* (New York: Viking, 1988), 24–25 and 281–82.

15. John C. Greene, *Science, Ideology and World View* (Berkeley: University of California Press, 1981), 7.

16. The meaning of Blumenbach's "Bildungstrieb" has been controversially debated. See, among others, Timothy Lenoir, "Kant, Blumenbach, and Vital Materialism in German Biology," *Isis* 71 (1980): 77–108; Robert J. Richards, "Kant and Blumenbach on the *Bildungstrieb*: A Historical Misunderstanding," *Studies in History and Philosophy of Biological and Biomedical Sciences* 31 (2000): 11–32; and John Zammito, "The Lenoir Thesis Revisited: Blumenbach and Kant," *Studies in History and Philosophy of Science, Part C: Biological and Biomedical Sciences* 43 (2012): 120–32. Further to theories of the origin of life and species, late eighteenth and early nineteenth century, see Nicolaas Rupke, "The Origin of Species from Linnaeus to Darwin," in *Aurora Torealis: Studies in the History of Science and Ideas in Honor of Tore Frängsmyr*, ed. Marco Beretta, Karl Grandin, and Svante Lindqvist (Sagamore Beach, MA: Science History Publications, 2008), 73–87.

17. Contrary to widespread belief, Haeckel did not synthesize the morphological approach with Darwin's theory of evolution by natural selection; he merely juxtaposed them, whenever opportune.

18. This thesis has been vigorously contested. For a lucid and spirited debate between Robert J. Richards (Darwin was substantively influenced by German Romantic biology) and Michael Ruse (Darwin was purebred British), see their jointly written *Debating Darwin* (Chicago: University of Chicago Press, 2016).

19. Other pros and cons about the Richards's thesis can be, and have been, adduced, but a discussion of them would carry us beyond the scope of this chapter. While there are many studies of the influences on Darwin in formulating his theory of evolution, not a single substantive study exists of Humboldt and the evolutionary ideas in his writings. A new exegesis of Humboldt's oeuvre in terms of "progressive development" is in preparation: Nicolaas Rupke, "Structuralist Evolution from Kant to Humboldt."

20. M. Hanna, *The Mobilization of Intellect: French Scholars and Writers during the Great War* (Cambridge, MA: Harvard University Press, 1996).

21. Georg F. Nicolai, *Die Biologie des Krieges: Betrachtungen eines Naturforschers den Deutschen zur Besinnung* (Zürich: Orell Füssli, 1817).

22. Louis-Félix Henneguy, "L'Allemagne et les sciences biologiques," in *Les Allemands et la Science* (Paris: Félix Alcan, 1916), 205–17, quote on 215 and 217.

23. Pierre Duhem, *German Science* (La Salle, IL: Open Court, 1991), 39.

24. See, among others, Robert Fox, *Science without Frontiers: Cosmopolitanism and*

National Interests in the World of Learning, 1870–1940 (Corvallis: Oregon State University Press, 2016); on the decline of German as a lingua franca, see also Michael D. Gordin, *Scientific Babel: How Science Was Done Before and After Global English* (Chicago: University of Chicago Press, 2015).

25. H. A. Turner, ed., *Hitler: Memoirs of a Confidant* (New Haven, CT: Yale University Press, 1978), 312–19 and throughout.

26. Heinz Brücher with Karl Astel, *Ernst Haeckels Bluts- und Geistes-Erbe* (Munich: Lehmanns, 1936).

27. Oberkommando der Wehrmacht, ed., *Weg zur Reifeprüfung. 5. Teil. Naturwissenschaften. Biologie—Chemie—Physik* (Breslau: Ferdinand Hirt, 1943), 97–98 (author's translation).

28. Houston S. Chamberlain, *The Foundations of the Nineteenth Century*, vol. 1, trans. John Lees (London: John Lane, 1911), 45.

29. Houston S. Chamberlain, *Immanuel Kant: Die Persönlichkeit als Einführung in das Werk* (Munich: Bruckmann, 1921), 522 (author's translation).

30. Nicolaas Rupke, *Alexander von Humboldt: A Metabiography* (Chicago: University of Chicago Press, 2007), 90.

31. For a study of Troll's plant morphology work and his structuralist leanings—yet that makes no mention of his Nazi sympathies—see Regine Classen-Bockhoff, "Plant Morphology: The Historic Concepts of Wilhelm Troll, Walter Zimmermann and Agnes Arber," *Annals of Botany* 88 (2001): 1153–72. Wilhelm Troll's brother, the distinguished geographer Carl Troll, opposed the Nazis (to his own personal risk) and was important in reestablishing geography and in ridding the subject of its associations with Nazi geopolitics in Germany after 1945.

32. Karl Beurlen, *Geologie und Paläontologie* (Heidelberg: Carl Winter, 1943), 10–11. See also Olivier C. Rieppel, "Karl Beurlen (1901–1985), Nature Mysticism, and Aryan Palaeontology," *Journal of the History of Biology* 45 (2012): 253–99.

33. Ute Deichman, *Biologists under Hitler* (Cambridge, MA: Harvard University Press, 1996), 74–80.

34. Uwe Hossfeld, *Gerhard Heberer (1901–1973): Sein Beitrag zur Biologie im 20. Jahrhundert* (Berlin: Verlag für Wissenschaft und Bildung, 1997). On the Ernst-Haeckel-Gesellschaft in Jena, see Uwe Hossfeld, *Geschichte der biologischen Anthropologie in Deutschland: Von den Anfängen bis in die Nachkriegszeit* (Stuttgart: Steiner, 2016), 273–83. A quadriga is a chariot, drawn by four horses abreast; a symbol of classical military prowess, it was adopted in this context with eugenicist overtones of racial conquest.

35. Otto Reche, "Die Genetik der Rassenbildung beim Menschen," in *Die Evolution der Organismen: Ergebnisse und Probleme der Abstammungslehre*, ed. Gerhard Heberer (Jena: Gustav Fischer, 1943), 683–706, quote on 705.

36. Heberer, *Evolution der Organismen*, iii.

37. Heberer, *Evolution der Organismen*, v.

38. This practice has persisted for many decades; see, for example, Richard Dawkins, *Climbing Mount Improbable* (London: Penguin, 1996), 208.

39. Stephen Jay Gould, foreword to Otto H. Schindewolf, *Basic Questions in Paleontology: Geologic Time, Organic Evolution, and Biological Systematics* (Chicago: University of Chicago Press, 1993), xi.

40. This information is based on personal conversations and correspondence with Fischer and with Mayr.

41. Georg Schneider, *Die Evolution das Grundproblem der modernen Biologie: Ein Abriß des Entwicklungsgedanken von Kaspar Friedrich Wolff über Darwin bis Lyssenko* (Berlin: Deutscher Bauernverlag, 1950).

42. I warmly thank the editors for their constructive criticism and substantive improvement of the form and content of this chapter, and Uwe Hossfeld and Georgy Levit at the University of Jena for helpful discussions about "alternative evolution theories."

PART THREE

GLOBAL STUDIES

CHAPTER SIX

Science, Sites, and Situated Practice

Debating the Prime Meridian in the International Geographical Congress, 1871–1904

CHARLES W. J. WITHERS

Reviewing in 2011 two edited collections on the history of science, philosopher of science Nicholas Jardine sketched what he considered "different stages towards understanding how science happened." Tracing the varying focus of historians of science upon triumphal tales of scientific progress, biographical (cum-hagiographical) treatments of "great geniuses" and canonical texts, and the rise of social constructivism and Actor Network Theory, Jardine discerned in more recent work a general rejection of grand narratives. "In accord with this decentring," he observed, "we see ever-increasing attention to local and indigenous styles of enquiry, culminating in the current predilection for 'geographies of science.' The focus here is on the conditioning of the practices and contents of the sciences by their national, institutional and geographical locations, and on the ways in which scientific knowledge has been distributed and mediated between sites of inquiry."[1]

There is not space here to debate more fully the substance and wider implications of Jardine's observations. But his suggestion that long-established concerns with *what* science was and by *whom* it was undertaken are now more directed at knowing *how* and *where* science was produced and received is surely an argument about enrichment, not essentialism, in the historical study of science. That we readily talk of the geographies of science and understand by it scrutiny both of the placed nature of science's discovery and justification and of its mobility over space owes much to the work of David Livingstone and others in addressing science's spatial dimensions.[2] Livingstone's has been a leading voice in insisting upon the locational significance of science's conduct and reception, and upon the connections among site, cognitive content, scientific performance, and social practice, whether of Darwin's work and reactions to Darwinism, John Tyndall's materialism, or of nineteenth-century scientific enquiries into climate and race.[3] Science must take place somewhere, of course: as Livingstone has noted, "location, like embodiment or temporality, is essential to knowing." Yet, as he has further observed, "The real question is, how do particular spaces matter in the production, consumption and circulation of science? In what ways do local and global forces conjoin to shape scientific culture in specific places? At what scale of analysis is the delivery of an identifiable set of claims to be apprehended?"[4]

This chapter considers these issues in examining a matter of global regulation locally articulated: the issue of a single prime meridian for the world as it was discussed in the meetings of the International Geographical Congress (IGC) between the first, in Antwerp in 1871, and the eighth, in Washington in 1904. My first concern is to address the ways in which the prime meridian question, a matter of global importance, was debated in the socio-scientific setting of an emergent international geographical body, to show by whom it was debated, how, and to what end. My second is to consider the implications of investigating the making of science as situated (in place) when it was also dated (in time). Rather than focus alone on the outcomes of international meetings, we may ask what it means to examine the epistemic and organizational practices that sustained them, as venues and as events within a wider and longer-run intellectual context. In Antwerp and elsewhere, delegates to the IGC, concerned as they were with a matter of global importance, were also constrained by the strictures of IGC protocol in what they could say about it, and even more constrained in what they could do about making their discussions effective. My argument is not that IGC meetings were in any definitive sense what Thomas Gieryn calls urban "truth spots" in establishing Greenwich as the world's prime meridian—that more complex story has been told elsewhere.[5] It is to observe that if the connections among science, speech, and site are at once placed and dated, of global import yet locally shaped, we need to show more exactly how, by whom, and why the science in question took the forms it did where and when it did. This may require that a balance be struck between analyzing site-specific moments and practices and synthesizing longer-run developments shaped in and by particular scientific venues. As others have argued, to attend only to the minutiae of science's making runs the risk of valorizing localist perspectives to the neglect of larger narratives within science's "spatial turn."[6] How, then, should we integrate local imperatives with global concerns when analyzing over time the placed and dated nature of science's conduct at different geographical scales?

The Prime Meridian and the Internationalization of Science and Geography after 1871

The prime meridian is the line and the point at which the world's longitude is set at 0°. Both longitude and time's measurement are based on the prime meridian. Since 1884, as one outcome of the Washington Meridian Conference, the world has had only one prime meridian: the Royal Observatory, Greenwich, in the United Kingdom. But before 1884, and even after, different prime meridians were at work in different places and for different purposes. In the ancient world, classical geographers and some Islamic scholars took the prime meridian to be Ferro (sometimes "Fero," in Spanish "Hierro"), the westernmost of the Canaries. The French formally acknowledged this position by royal edict in 1634 but also reckoned their prime meridian from Paris. The Dutch based their prime meridian on the Canaries but from Tenerife not Ferro. The British took their prime meridian either from different islands in the Azores or from London (by which was usually meant, before about 1770, St. Paul's Cathedral, not Greenwich). From 1850 to 1912, the United States employed two prime meridians: Washington for topographical and astronomical purposes, and Greenwich for maritime navigation, and, after 1883, for the regulation of its railways.

In short, by the late nineteenth century, there was no single point of calculative origin from which to measure the world's space and time. Furthermore, different metrological standards were at work: the imperial and the metric. What distinguished discussion about the prime meridian in the settings of the international geographical congresses held from 1871 was its overtly international tenor. This can be explained by the ways in which science was increasingly "internationalized" during the late nineteenth century. For Elisabeth Crawford, "the universe of international science" in the half century from 1880 was distinguished by three features: cognitive homogeneity (the sharing of common problems and methods within and among discrete disciplines), standardized communication (with international associations and journals increasingly crossing national and disciplinary borders), and new agreements over technical standards—in advocating the metric system, for example—and in an emerging consensus over the procedures for scientific research in laboratories and in universities.[7]

The establishment of the IGC was part of this internationalization of science in the late nineteenth century and was so from concerns to promote the discipline of geography. The 1871 meeting—the world's first international geographical meeting—held August 14–22, 1871, in Antwerp, was envisioned as a new beginning for the sciences of the earth. Charles d'Hane-Steenhuyse, vice president of the Organising Committee, intended that the gathering should "arouse the geo-cosmographical sciences from their long sleep."[8] Geography was then not the professionalized and institutionalized subject it would later become. Certainly, it was widely taught in schools, in universities, and in military academies. There was a well-established geographical publishing industry. Geographical exploration commanded widespread public interest and extended the geopolitical reach of European nations in particular. National geographical societies appeared in the nineteenth century to direct and report upon these activities: twenty-two such societies were founded before 1871 and thirty-nine in the decade after 1871.[9]

But professional geographers were few and far between. University departments of geography appeared only late in the nineteenth century. Geography was not everywhere agreed upon as a science, nor yet international in its procedures and methods. For these reasons, we should not see debates in the IGC between 1871 and 1904 on the need for a single prime meridian as the associational outcomes of a subject whose intellectual definition was certain and whose practitioners shared common problems in standard ways. Discussions about a single prime meridian and its global utility were, I argue, not a consequence of geography's established disciplinary identity, cognitive homogeneity, and shared technical standards; rather, the reverse was true. That is, debates over a single global prime meridian in late nineteenth-century international meetings helped produce geography's emergent status as an international science.

Internationalizing Geography as a Global Subject: Debating the Prime Meridian in the IGC

Detailed study of the IGC is a major lacuna in the history of nineteenth-century geographical knowledge. Early IGC events were commercial gatherings, trade exhibitions, and academic meetings: manufacturers of geographical apparatus, cartographic firms, globe makers, and geographical publishers were all present, yet this dimension of geography's history has been overlooked. Manuscript records of the IGC are very few. We know by subject the content of the papers delivered at the meetings, but the published printed records (*Comptes Rendu*) do not present a full account of the spoken proceedings. As Charles d'Hane-Steenehuyse had intended, the first meeting, in Antwerp in 1871, "served to stimulate the already reviving interest in the study of geography throughout Europe."[10]

This reviving interest was managed in several ways. At each IGC meeting, the executive committee—made up, usually, of leading geographers and civic dignitaries from the host city—met with geographers and scientists from other nations in order to identify possible future venues. No formal record exists of decisions over competing candidate cities or nations. Future meetings appear to have been arranged at the invitation of national geographical societies, rather than governments, and often several years in advance. Yet there was an element of competition between host sites: when Gardiner Hubbard, president of the National [American] Geographic Society, "sent a formal invitation to the Sixth International Geographic Congress [in London in 1895] inviting the congress to hold its seventh meeting in the United States," his invitation was rejected in favor "of the geographers of Germany," who had offered to host the seventh meeting in Berlin.[11]

Surviving records make clear the concern from the outset to implement procedures and rules of conduct by which to manage IGC business. In order to coordinate academic discussion, the IGC organized itself into parallel subject-based groups that focused upon areas of delegates' common interest. There was some shift in groups' subject content within the IGC. New subject-based groups were added as the IGC grew in size. Issues raised in individual groups, or even by individuals within groups, were brought forward to a general IGC session where delegates would vote upon specific group-derived questions. This general session became the executive forum or congress of the IGC. That is, IGC business was thematically structured and had to be voted upon in order to proceed to further discussion, and there were strict procedures for identifying subjects of shared significance across groups.

Preliminary Concerns about Global Regulation: Antwerp 1871

In Antwerp, the prime meridian was the focus of group I, devoted to cosmography, navigation, and international commerce. Its chair was the British naval officer and Arctic explorer Vice Admiral Erasmus Ommaney, a leading figure within the British Association for the Advancement of Science and a council member of the Royal Geographical Society in London. The group's main concern was uniformity in cartography (or, rather, its lack): in mapping scales and in symbolization. The need for a single initial meridian was part of these issues. Under Ommaney's direction, the question of a common first meridian came before delegates six times in group discussion during the Antwerp meeting.[12]

Ommaney presented delegates with a simple question: "Could we not agree to adopt the same prime meridian?" For some delegates, the answer lay in the advantages to science (their responses emphasized issues of accuracy and error). Others stressed matters of daily practicality (common usage and convenience/inconvenience) or the need for cartographic and metrological standardization. Delegates understood that having different prime meridians was awkward: the need for metrological uniformity was paramount given the constant need to calculate and then recalibrate the longitudinal and astronomical differences among different originating base points. Yet there was no common agreement within group I on the issue. Reporting to the Royal Geographical Society on the Antwerp meeting, Ommaney noted that "several questions of international importance" had been dealt with there, including "the possibility of adopting the same first meridian by all nations," but neither he nor the society's council took the matter further.[13]

Closer examination reveals national and subject-based fracture lines within this international setting. In Antwerp, the proposal was made that Greenwich be adopted as the world's initial meridian. Other delegates, notably the French, argued that the Paris prime meridian should be the world's 0° baseline since it was also widely used in science and had a claim of primacy over Greenwich as an observed prime meridian given the establishment of the Paris Observatory in 1667. For French hydrographer Adrien Germain, secretary to group I in Antwerp, Paris was the obvious candidate for global adoption, not Greenwich, given the Paris Observatory's historical importance and because, for the French, *le meridién* based on Paris was employed as an observed prime meridian, on maritime charts, and in the nation's topographical mapping and geographical teaching. The fact that this view was not shared by all French delegates reinforces the sense that differences expressed within the IGC were apparent on disciplinary grounds, even as a matter of individual preference, rather than simply as a matter of national allegiance to any one prime meridian. To Émile Levasseur, Germain's French geographical colleague in Antwerp, the precedence of Paris was undeniable, but the choice—on pragmatic grounds—had to be Greenwich for the good, practical reason that the great majority of extant nautical charts used Greenwich. Although there was international concern over the need for a single global point of origin from which to measure the world, the proposed solutions reflected national and subject-based positions. The result was that the proposal finally arrived at was limited in reach and, in its wording, always likely to be limited in effect: "The Congress expresses the opinion that for maritime routing charts [i.e., pilot or navigational charts], a first meridian be adopted, that of the Greenwich Observatory; and that after a period of time, say in ten to fifteen years, this initial position is to be made absolutely obligatory of all charts of this nature."[14]

This resolution, which was voted upon and accepted in IGC general session, is the first expression of conjoint international resolution over a single initial prime meridian for the world. But because it concerned only one sort of map employed by one user community only, it was never likely to be formally ratified. The IGC had no authority to insist that governments or scientific users accept its recommendations. The proposal left geographers and topographical surveyors using different 0° start points for their maps. Astronomers continued to use the observed prime meridian relative to their observatory. The status quo ante was maintained: differences in practice existed between different users and among adherents to different prime meridians.[15]

The Debate Continued: Paris 1875

The second IGC was held in Paris in August 1875. The local organizer was Adrien Germain, who, in Antwerp, had declared against Greenwich and in favor of the Paris prime meridian as the world's prime meridian. To Germain, no country should have to adopt a neutral shared meridian, France least of all. In Paris, Germain used his organizational role to manage the meeting's business. In acting thus, Germain set a tone for the Paris meeting that, in group sessions and in congress-level discussions, sought to limit collective debate over a single prime meridian for fear that Greenwich might prevail.[16]

In the specialist groups—the key academic debating spaces of the IGC—the question was discussed by delegates in two groups. Group II—Hydrography and Maritime Geography—debated the issues of cartographic uniformity, the predominance of Greenwich in maritime charts, and the proposal adopted in Antwerp but only to reiterate the earlier view that a common first meridian should be adopted. Delegates to group I—Mathematical Geography, Geodesy, and Topography—took the issue further. For them, the choice of a shared initial meridian could not be separated from questions of a uniform metrology, universal time, and the common good of a shared baseline for the world's measurement. Alexandre-Émile Béguyer de Chancourtois, the French superintendent of mines, favored a prime meridian that ran through the Azores on the grounds that such a line separated the Old World from the New and that a mid-Atlantic position made a good baseline against which to change the civil day. Others argued for a prime meridian in the Pacific, between the Asian and American continents, at 180° from either Paris or Greenwich.[17]

Britain's geographical representatives were less involved in Paris in 1875 than Ommaney had been in Antwerp in 1871 because of the Royal Geographical Society's rather haughty attitude toward the Paris IGC. Late in 1874, Clements Markham, president of the Royal Geographical Society, had requested that Britain's foreign secretary, Lord Arthur Russell, write to his French counterpart to advise him "that at present the English [*sic*] Government show no inclination to take any official part in the Geographical Congress & Exhibition as not being held by desire or under the Authority of the French Government, but as the action of a private Society." There was no difficulty with the two-part nature of the IGC but, seemingly, every difficulty with it being organized as a nongovernmental event.[18]

In Paris, a representative of the Geographical Society of Geneva argued that the world's prime meridian should center upon Jerusalem. The point of this suggestion was that Jerusalem was "neutral": no country could object to siting the world's baseline there. Opponents argued that the initial meridian could not be sited there because there was no astronomical observatory against which to position an observed meridian. This view was simply countered: the line should be determined first as a statement of universal principles, with an astronomical observatory built afterward to affirmation this collective enterprise. Other delegates argued for Ferro.

Antwerp began discussion over the internationalization of the prime meridian in formal geographical meetings yet ended in a weak and limited proposal. Debate in Paris highlighted persistent national differences, with the addition of the Jerusalem proposal adding to the confusion. Yet it is possible to see in the discussions in Paris further shared recognition of the virtues of a single meridian. Sectoral differences within the term "scientific community" were again clearly apparent. Broadly, astronomers were content with different national observed prime meridians and with the calculations this demanded. Some, such as French astronomer Antoine François Joseph Yvon-Villarceau, saw merit in the universal acceptance of a common initial meridian provided related amendments were made to the reckoning of time. Metrologists emphasized the virtues of a common initial meridian for topographical maps. Maritime communities looked commonly to Greenwich, as they had done in Antwerp, but Greenwich was not strongly endorsed by others. Geographers and cartographers continued to employ different measured prime meridians.

At Paris, participants in group I were sufficiently convinced of the merits of a common initial meridian to vote upon the issue, but in one context only—strictly with reference to its use in world maps and atlases. The prime meridian they chose, and by a large majority, was Ferro. This was taken to be 20° west of Paris. Delegates proposed that this point should be described by all atlas publishers as "the meridian of origin." The IGC congress did not accept this recommendation. This was, possibly, a direct result of Germain's influence, because French delegates outnumbered those from other countries; because British voices were relatively silent, believing the issue to have turned in Greenwich's favor; or because others recognized that Paris carried considerable legitimacy as a world-leading prime meridian.[19]

Post-Paris Concerns over a Prime Meridian, 1875–1881

Reporting to colleagues in Geneva, the Swiss delegation to Paris identified three key stumbling blocks to common agreement upon a single prime meridian. The first was "multiplicity"—there were too many different prime meridians in use. The second was the "confusion" caused by adherence to different initial meridians for different purposes. The third was "inertia." This was a twofold problem: the lack of agreement over which prime meridian should be used for which specific purpose or which prime meridian for all purposes.[20]

Following Paris, Swiss delegate Henri Bouthillier de Beaumont, president of the Geographical Society of Geneva, declared himself in favor of a prime meridian that ran through the Bering Strait, 150° west of Ferro. The extension of this meridian over both poles would, de Beaumont argued, form a meridian arc passing through numerous European countries and the African continent but not through any nation's capital or through an astronomical observatory. This positioning for what he called the "*médiateur*"—perhaps in reference to the equator (in French, *equateur*), or because his proposition diplomatically mediated between competing options—was, he continued, the best solution for a global prime meridian that was everyone's and no one country's.[21] The June 1879 edition of *Popular Science Monthly* lauded the term "mediator" in its summary of the issue, the author declaring himself in favor of Bouthillier de Beaumont's "project," proclaiming its use to be "the common property of all the civilized nations, . . . neutral ground, a position independent of all political power, and under guarantee of all the states of the civilized world."[22]

By this point, five different prime meridians were under discussion in the IGC as candidates for the world's single baseline—Paris, Greenwich, Jerusalem, Ferro, and one in the Bering Strait. The prime meridian question was not aired at the 1876 Brussels IGC, which focused on the division of sub-Saharan Africa by Europe's imperial powers. When the topic was next raised, at the International Congress of Commercial Geography in 1878 in Paris, a resolution was passed there, drawing attention to the inconvenience of having numerous prime meridians, but no solution was offered. In the 1879 meeting of this body, in Brussels, delegates again recognized the collective benefits of a common shared meridian (many of them proposing Greenwich), but, after debate, they reiterated the Antwerp resolution.[23]

De Beaumont's post-Paris proposal for a Bering Strait prime meridian, 150° from Ferro, had its supporters. Charles Daly, mayor of New York City and president of the city's Geographical Society, considered de Beaumont's plan a solution to persistent national differences. Appraising developments in geography in an 1879 review, Daly reported that "several attempts have been made during the past few years to get the different nations of the world to agree upon a common meridian, instead of having each adhering to its own, like the meridians of Greenwich, Paris, Washington, &c." Since "the meridian of Greenwich is the one most extensively found on maps and charts," Daly observed how Americans have "generally been disposed to adhere to it, and would probably be quite willing, as a nation, to unite in its adoption." But, he continued, when the question of doing so had arisen at the Antwerp IGC, "the disposition of the French members was to adhere to the meridian of Paris, and as it seems to be difficult to get one nation to adopt the meridian of any other, the object might be effected by adopting a common meridian solely upon geographical grounds."[24]

Daly reckoned Bouthillier de Beaumont's Bering Strait proposal sensible for several reasons: it could "be very easily connected" (in astronomical and mathematical calculations) with other "principal meridians," and it would divide Europe into east and west, "thus giving a division which has been tacitly recognized for ages." Passing as it would through many nation states, "it would become really an international meridian, as each nation might establish a station or observatory on the line of it." It was, in short, globally expedient: "As this would be, for the reasons above suggested, a very desirable first meridian, and as there appears to be no other way of getting over the disposition of nations to adhere to their own and of avoiding the confusion of having so many, I fully concur in M. de Beaumont's suggestion, and hope, as a practical relief from an existing difficulty, that it may hereafter be generally adopted."[25]

This evidence is of interest given the stated international support for a candidate prime meridian other than the one finally agreed upon at the 1884 Washington Meridian Conference (Greenwich). Understanding Crawford's "universe of international science" depends on exploring different institutional sites and disciplinary practices, at a moment and over time. But the evidence also demonstrates the "travel" of the issue out from IGC meetings as well as its contested making within the meetings. That it does so highlights the difficulties that the mobility and diversity of scientific ideas presents to modern scholars as they endeavor to understand the articulation of the question in, but also beyond, institutional settings that were, at one and the same time, sites of science's making and of its reception.

Moves toward a Resolution: The 1881 Venice IGC

At the third IGC, organized by the Italian Geographical Society and held in Venice in September 1881, the prime meridian was the principal business of group I. Unlike in Antwerp and Paris, the Venice IGC was attended by delegates from North America representing organizations that had not hitherto been formally involved in IGC discussions but whose interests centered upon global metrology: the American Metrological Society, the US Army Corps of Engineers, and the American Geographical Society.

George Wheeler of the US Army Corps of Engineers drew attention to the confused state of affairs over the world's prime meridians that delegates in Venice had met to remedy:

> Upon examining specimens of the extended general topographical map series of Europe fourteen separate and independent meridians of reference are found: (1) Greenwich, for the United Kingdom and India; (2) Paris, for France, Algeria, and Switzerland; (3) Lisbon, for Portugal; (4) Rome, for Italy; (5) Amsterdam, for Holland; (6) Isle of Ferro, westernmost of the Canaries, for Prussia, Saxony, Wurtemberg, and Austria; (7) Ferro and Christiana [Oslo], for Norway; (8) Copenhagen, for Denmark; (9) Madrid, for Spain; (10) Stockholm and Ferro, for Sweden; (11 and 12) Ferro, Pulkowa,

> Warsaw, and Paris, for Russia; (13) Brussels, for Belgium; and (14) Munich, for Bavaria.
>
> Both Greenwich and Washington, principally the former, have been used for maps of land areas in the United States.

In addition to these topographical cartographic prime meridians, Wheeler identified those in use in hydrography: "The meridian of Greenwich is used on the Government marine charts of England, and India, Prussia, Austria, Russia, Holland, Sweden and Norway, Denmark, and the United States; while France employs Paris; Spain, Cadiz; Portugal, Lisbon, and Naples is found on some Italian as well as likewise Pulkowa on certain Russian hydrographic charts." Further, noted Wheeler, "as nautical and astronomical tables came into more general use a number of meridians of reference were established, as at Toledo, Cracow, Uranibourg [*sic*], Copenhagen, Goes [*sic*], Pisa, Nuremberg, Augsburg, London, Paris, Rome, Greenwich, Washington, Vienna, Ulm, Berlin, Tubingen, Venice, Bologna, Rouen, Dantzig, Stockholm, St. Petersburg, &c."[26]

Debate in Venice on these national and thematic differences centered around three spoken papers. "Paper A" was given by Charles Daly of the American Geographical Society on behalf of his fellow American Frederick A. P. Barnard of the American Metrological Society. The Barnard-Daly paper argued that the prime meridian to be used was "the meridian situated in longitude one hundred and eighty degrees, or twelve hours distant from the meridian of Greenwich . . . which meridian passes near Behrings [*sic*] Straits and lies almost wholly on the ocean." This view was shared by General William B. Hazen, also of the American Metrological Society. His paper, "Paper B," was given on his behalf by George Wheeler. "Paper C" was given by Sandford Fleming, who spoke on behalf of the Royal Canadian Institute of Science in Toronto and the American Metrological Society. In 1884, Fleming would be one of Britain's four-man delegation to the Washington Meridian Conference.

Fleming argued for "the establishment of a Prime Meridian and Time-zero, to be common to all nations." Fleming's argument revolved not around the many prime meridians then in use but upon "the relations of time and longitude and the rapidly growing necessity in this age for reform in time-keeping." Standardization of what he called "Cosmopolitan Time" was his and others' end in view, but "the first step towards its introduction is the selection of an initial meridian for the world."[27]

The papers and resolutions presented at the Venice IGC focused on two proposed initial meridians, either the Bering Strait *médiateur* (150° from Ferro) or the Greenwich anti-meridian (180° from the Greenwich Observatory), and on the prime meridian's connections with universal time. Again, it is important to understand how the IGC worked in order to explain the cognitive outcomes. The virtues of a single prime meridian were recognized. Yet no one prime meridian was agreed upon. Fleming proposed Greenwich's anti-meridian. Others in group I held fast to Ferro. Other options were advanced: Béguyer de Chancourtois, who in Paris in 1875 had proposed the Azores, in Venice offered several different prime meridians, urging universal acceptance of the metric system as he did so.

Given these differences, group I brought forward what was only a limited proposal: "Group I expresses the hope that within a year Governments appoint an international commission for the purpose of considering the subject of an initial meridian, taking into account the question of longitude, but especially that of hours and dates. . . . The President of the Italian Geographical Society is requested to undertake the steps necessary to realize this view via his Government and the foreign geographical societies." Because Washington was mentioned as a venue for the proposed meeting of an international commission, it is tempting to regard the 1881 Venice IGC as a formative precursor of the recommendations arrived at in Washington in 1884. It is clear that, in Venice, concern was expressed that a commission be established "for the purpose of considering the subject of an initial meridian," a commission "which should be composed of scientific members, such as geodesists and geographers and of persons representing the interests of commerce, learning, &c."[28] But that is to place an unwarranted emphasis upon later "success" to the neglect of considering the several expressions of "failure" in the several IGC meetings. Delegates in Venice could not agree among themselves, and what limited resolution was advanced was not supported by the Venice IGC congress as a whole: the resolutions that remained were the limited ones adopted in Antwerp and in Paris.[29]

Continued Uncertainty: The Prime Meridian in Geographical Congresses, 1891–1904

Among the audience at the fifth IGC, in Berne in 1891, was Thomas Holdich, the British military surveyor and geographer (and, later, president of the Royal Geographical Society). In his report to the Royal Geographical Society upon the Berne meeting, Holdich made clear the continuing indecision over the prime meridian:

> The meridian question, although it is apparently as far from solution as it was previously to the Washington Congress, has certainly advanced far enough that all English [*sic*] maps should possess a common origin for longitude. At present this is not so, for maps of India and of parts of the bordering countries are published with a longitude value based on an incorrect assumption of the position of the Madras observatory, differing about two and half miles from the true Greenwich value; so that, as our mapping extends westward through Persia and eastwards through Burma we become involved in awkward discrepancies. I would venture to suggest that the opinion of the Surveyor-General of India should be consulted as to the advisability of adopting the Greenwich meridian in future for all Indian mapping. I am quite aware of the nature of the reasons which have prevented its adoption hitherto, but since attending this Congress I have come to the conclusion that a continuance of the present system is a grave disadvantage if we wish to persuade other nations to adopt Greenwich as their longitude of origin, and that this disadvantage outweighs previous considerations.[30]

In addition to the multiplicity of prime meridians on British imperial topographic maps, the maps produced by other countries demonstrated the continued use of different prime meridians even after the 1884 Washington meeting had proposed Greenwich as the world's 0° baseline. In 1885, fifteen different national prime meridians were employed on large-scale topographic mapping throughout the world. By 1898, all but one were still in use for national topographic purposes (only Warsaw, used as a prime meridian in some Russian topographic mapping, had been dropped in the interim).[31]

This problem and the prevalence of different scales and standards in national topographic mapping was addressed in Berne in 1891 by German geographer Albrecht Penck, who announced there his scheme of world mapping at a scale of 1:1,000,000 (1:1M). While scholars of the so-called Millionth Map Project have shown its importance, they have overlooked its significance with regard to wider debates in the internationalization of science, the IGC, and the prime meridian. Penck's scheme was linked to a resolution passed in Berne that the "English prime meridian" (Greenwich) should be universally adopted (delegates who proposed this did so in the hope that Britain should adopt the metric system).[32] Penck returned to his scheme at the sixth IGC, in London in 1895. So, too, but in absentia, did Bouthillier de Beaumont. Other contributions included a talk from Enrico Frassi of the Italian Geographical Society on time reform and a system of hour zones.[33]

At the seventh IGC, held in Berlin in 1899, it was unanimously recommended, in relation to Penck's global cartographic project, that Greenwich be used as the world's longitudinal base 0° and that the meter be employed in its topographical measurement and cartographic representation. This recommendation was not new: it had first come forward in 1895. In a paper in Berlin on the advantages to geography of the metric system, British geographer Hugh Robert Mill shed light on the discussions at the earlier London IGC. There, French delegates had proposed that heights on Penck's proposed 1:1,000,000 world map should be given in meters, and that the world's prime meridian should be Greenwich. This "very happy international compromise . . . might before long be extended to all maps and geographical writings," Mill noted. After all, "The objection to the metric system in English-speaking countries is no more strongly based than the objection to accepting the meridian of a foreign observatory as the zero of longitude is in France. The vast benefit of international uniformity in standards should outweigh all other sentiments."[34]

By 1904, the IGC had taken heed of Mill's views. At the Washington IGC that year—twenty years after Greenwich's status in this respect had been agreed—the IGC affirmed Greenwich as the world's prime meridian in relation to time: "In view of the fact that a large majority of the nations of the world have already adopted systems of standard time based upon the meridian of Greenwich as prime meridian, . . . this congress is in favour of the universal adoption of the meridian of Greenwich as the basis of all systems of standard time."[35]

Siting and Dating Science in Practice

In his 2007 review of the historical geographies of science, Richard Powell pointed to the emergence of ethnographic studies of laboratory practice as a distinct feature within work on the sociological and geographical study of science, and as particularly illustrative of the "localist turn" and of the importance of practice. Powell's remark that "it is incumbent upon analysts to investigate what counts as scientific practice in *particular* cases" may be easier to achieve for science in contemporary settings than in historical context.[36] One contemporary ethnographic study of the cultures of classification by taxonomists whose purpose was to agree upon and accept biological DNA bar codes for species as part of the wider "genomics revolution" was based upon six years of anthropological enquiry within international meetings: observations of participants' practice, the rhetoric of appeal to unity in species identification, the protocols observed in conference sessions, and so on.[37]

Historians and geographers of science seeking to explain science's production, reception, and mobility in the nineteenth century do not have the luxury of prolonged firsthand ethnomethodological encounters with the ambiguities of science's making. They are, more commonly, confronted with the unequal archival traces of that making, at a moment, in specific settings, or in its varying expression over time and in different places. Printed records of the content of speeches and the structure of scientific meetings do reveal what was said, and so can illuminate the content of debate in relation to specific places as "platform science" or "speech spaces" or as scientific "shop talk."[38] They do not as readily disclose any procedural controls upon the performance of the science, or audience reaction either in situ or elsewhere later. Yet looking "internally," as it were, at how different institutions of science worked, at what was possible to do as well as say within that setting and what not, and not alone at their "external" consequences as if scientific bodies and events acted always of their own volition, might cast light on how more exactly scientific knowledge was made and received.[39]

The case of the prime meridian in different IGC meetings cannot be divorced from its wider intellectual and political context. In one sense, the meetings of the IGC could be taken as uneven, perhaps inconsequential. In the delegates' and groups' inability to agree among themselves over the advocacy of a single preferred candidate meridian despite collective agreement in principle and because the "solution" was advanced elsewhere at another time, debates in the IGC proved inconclusive, even unproductive. International debate in a local setting managed strictly in order to effect a global solution had its apotheosis in the Washington Meridian Conference in 1884: the IGC both foreshadowed this and postdated it, with earlier meetings failing to get beyond limited resolutions and later IGC meetings seemingly oblivious of the Washington resolutions. The early IGC meetings, notably those in Antwerp (1871), Paris (1875), and in Venice (1881), shaped particular but limited resolutions regarding a single prime meridian. Later IGC meetings—those in Paris (1889), Berne (1891), London (1895), Berlin (1899), and Washington (1904)—did not recognize in their formal business that, in Washington in 1884, others had met to propose and agree upon Greenwich as the world's initial meridian. What study of these IGC meetings does reveal, however, is that this proposal was not enacted upon equally—as Holdich made clear in recounting his sense of the 1891 Berne meeting and as Mill and Penck and others would testify from other post-1884 IGC meetings.

This complex picture of the IGC and the geographies of the prime meridian has wider implications. Fuller pictures of science's making and reception depend upon our knowing where else given topics were articulated other than in successful sites of discovery or justification, in showing why things elsewhere, there and then, acted to limit or curtail debate. As science sought to internationalize in the later nineteenth century, associational culture within and among disciplines can be seen as something produced: it was not innate to emergent disciplines. Scientific meetings with an annual or other regular cycle of occurrence allow for the possibility that given topics can be traced over time, in different places but within the same institutional confines. This is more applicable to the nineteenth century than for earlier periods when the emphasis upon international and, even, upon disciplinary specific practices was less evident. Even as this is true, however, one consequence of working with the languages of "localness" and of place and with "the global"—certainly with their combination—may be the relative elision of "the national" as a focus of study. Has the presence of longer-distance networks, cycles of exchange, and accumulative regimes or their local, grounded manifestations meant that we are also witness to what one historian of science has called an "end to national science"?[40] Here, we must recognize that creative tension between "locality and globality" that Peter Galison identified as one of "ten problems in the history and philosophy of science" and understand that different sciences will have different histories and geographies in that respect.[41]

A counterargument would have it that the effect of using terms such as "internationalism" or "transnationalism" (even global) in relation to the scales at which we examine the histories and geographies of science has been to establish the nation even more strongly as a frame of reference. As the authors of one conjoint commentary upon the idea of a transnational history of science have observed: "Other approaches such as world history and new global history share with transnational history the wish to abandon Euro- and US-centric viewpoints, explaining the role of historical actors and agencies in international networks, and focusing on the circulation of people, objects and ideas. Transnational history approaches appear to differ from these perspectives because of their focus on the modern and contemporary periods, and the reappraisal of nations' role in shaping the past. Indeed, because of these distinctive chronological and theoretical features, some scholars have argued that transnational history pays more attention to nation states than do its alternatives."[42] Transnational history may not be a specific set of instruments for historical (or geographical) work so much as a way of framing questions such that comparisons, locally expressed, in and across different national contexts, come more to the fore.

The example of a single prime meridian for the world as it was debated in the meetings of the IGC from 1871 demonstrates an element of cognitive homogeneity in the internationalizing of nineteenth-century science and geography, at least by intention. At the same time, the evidence within different local sites of practice that shared the same ends in view—concerning geography as a science, the global common good, and over the prime meridian as an expression of both—suggests something much less than homogeneity. Different metrological systems were at work. Different views over technical standards—the utility of the metric system, for example—were used as elements of negotiation over different prime meridians. In IGC meetings, sectoral differences among geographers, astronomers, mapmakers, and navigators meant that no shared view could be advanced, even as it was recognized that such a thing would be beneficial. Such recommendations as were advanced in Antwerp in 1871, Paris in 1875, and Venice in 1881 reflected the interests of one user community but certainly not all. In Venice, for all that the meeting exposed regarding the interests of the American scientific community and others over the prime meridian, it was in the eyes of one delegate a success more for its commercial and trade elements than for its international academic character.[43]

Scientific meetings do not contain or restrict the subjects voiced or demonstrated there within institutional or disciplinary confines. That scientific ideas travel to be debated elsewhere and in the traveling may be modified is clear from study of Tyndall's 1874 address in Belfast or of the Wilberforce-Huxley debate in Oxford in 1860, to name but two instances.[44] If science's making is a process of "knowledge in transit" as Secord has it, or of "spaces of rhetoric" as Livingstone puts it, this requires that researchers look outside the meetings themselves to see the effects of scientific discourse upon public comprehension, and vice versa, and to show how, if they were, later events to the same purpose were shaped by discussion somewhere else in the interim.[45] This supposes that the topic is returned to at other specialist sites and times. Of the public understanding of science, it also supposes that the public has not been circumscribed from engagement with scientific topics for one reason or another. This might mean specialist languages or procedures being invoked, or a lack of cognitive homogeneity in the public sphere even though that nebulous concept "the public" may have its own tacit understanding of the issue in hand.[46]

Looked at over time, and in regard to the several IGC meetings, debates over the prime meridian as the world's 0° for the calculation of longitude and universal time varied by place and, within each IGC, by user community and by candidate prime meridian. The 1884 Washington Meridian Conference was a key moment and site toward regulation of the world's measurement notwithstanding the fact that the recommendations over Greenwich were not taken up equally by all nations. Study of the international geographical settings in which the question of a single prime meridian for the world was debated before and after 1884 reveals a complex geography to the emergent science of geography. It also highlights the need to consider, where and when we can, where else science was made and articulated other than in those venues and practices associated with its success.

Notes

1. Nicholas Jardine, "Chalk to Cheese: Progress, Power, Cooperation and Topography: Stages towards Understanding How Science Happened," *Times Literary Supplement* (December 16, 2011): 3–4, quote on 3.

2. Several studies chart the placed nature and geographical study of science: Adi Ophir and Stephen Shapin, "The Place of Knowledge: A Methodological Survey," *Science in Context* 4 (1991): 3–21; David N. Livingstone, *Putting Science in Its Place: Geographies of Scientific Knowledge* (Chicago: University of Chicago Press, 2003); Diarmid A. Finnegan, "The Spatial Turn: Geographical Approaches in the History of Science," *Journal of the History of Biology* 41 (2008): 369–88; Jan Golinski, *Making Natural Knowledge: Constructivism and the History of Science* (Cambridge: Cambridge University Press, 1998); Richard Powell, "Geographies of Science: Histories, Localities, Practices, Futures," *Progress in Human Geography* 31 (2007): 309–30; Crosbie Smith and Jon Agar, eds., *Making Space for Science: Territorial Themes in the Making of Knowledge* (Basingstoke: Macmillan, 1998); Charles W. J. Withers and David N. Livingstone, "Thinking Geographically about Nineteenth-Century Science," in *Geographies of Nineteenth-Century Science*, ed. David N. Livingstone and Charles W. J. Withers (Chicago: University of Chicago Press, 2011), 1–19.

3. From his extensive body of scholarly work, the following illustrate Livingstone's importance in shaping the subfield "geographies of science": David N. Livingstone, "Darwinism and Calvinism: The Belfast-Princeton Connection," *Isis* 83 (1992): 408–28; "The Spaces of Knowledge: Contributions towards a Historical Geography of Science," *Environment and Planning D: Society and Space* 13 (1995): 5–34; *Putting Science in Its Place*; "Science, Site and Speech: Scientific Knowledge and the Spaces of Rhetoric," *History of the Human Sciences* 20 (2007): 71–98; *Adam's Ancestors: Race, Religion and the Politics of Human Origins* (Baltimore: Johns Hopkins University Press, 2008); "Politics, Culture, and Human Origins: Geographies of Reading and Reputation in

Nineteenth-Century Science," in Livingstone and Withers, *Geographies of Nineteenth-Century Science*, 178–202; *Dealing with Darwin: Place, Politics, and Rhetoric in Religious Engagements with Evolution* (Baltimore: Johns Hopkins University Press, 2014); "Debating Darwin at the Cape," *Journal of Historical Geography* 52 (2016): 1–15.

4. David N. Livingstone, "Text, Talk and Testimony: Geographical Reflections on Scientific Habits. An Afterword," *British Journal for the History of Science* 38 (2005): 93–100, quote on 100.

5. For fuller discussion of the geographies of the prime meridian, see Charles W. J. Withers, *Zero Degrees: Geographies of the Prime Meridian* (Cambridge, MA: Harvard University Press, 2017), and, on the regulation of time in relation to the prime meridian, see Ian R. Bartky, *One Time Fits All: The Campaigns for Global Uniformity* (Stanford, CA: Stanford University Press, 2007). On Gieryn's work on "truth spots," see Thomas Gieryn, "Three Truth-Spots," *Journal of the History of the Behavioral Sciences* 38 (2002): 113–32, and "City as Truth-Spot: Laboratories and Field Sites in Urban Studies," *Social Studies of Science* 36 (2006): 5–38.

6. Steven J. Harris, "Long-Distance Corporations, Big Sciences, and the Geography of Knowledge," in *The Scientific Revolution as Narrative*, ed. Mario Biagioli and Steven J. Harris, special issue of *Configurations* 6 (1998): 269–305; Steven Shapin, "Placing the View from Nowhere: Historical and Sociological Problems in the Location of Science," *Transactions of the Institute of British Geographers* 23 (1998): 5–12; James A. Secord, "Knowledge in Transit," *Isis* 95 (2004): 654–72.

7. Elisabeth Crawford, "The Universe of International Science, 1880–1939," in *Solomon's House Revisited: The Organisation and Institutionalization of Science*, ed. Tore Frängsmyr (Canton, MA: Science History Publications, 1990), 251–69. M. Geyer and J. Paulman, eds., *The Mechanics of Internationalism: Culture, Society, and Politics from the 1840s to the First World War* (Oxford: Oxford University Press, 2001).

8. *Compte-Rendu du Congrès des Sciences Géographiques, Cosmographiques et Commerciales Tenu a Anvers du 14 au 22 Aout 1871. Tome Second* (Anvers: Gerrits and Van Merlen, 1872), 1:176.

9. The literature on the history of geography in these terms is voluminous. For this summary, I have made use of David Livingstone, *The Geographical Tradition: Episodes in the History of a Contested Enterprise* (Oxford: Blackwell, 1992); the essays in Gary S. Dunbar, ed., *Geography: Discipline, Profession and Subject since 1870* (Dordrecht: Kluwer Academic, 2000); Karin Morin, *Civic Discipline: Geography in America, 1860–1890* (Farnham: Ashgate, 2011); and Charles W. J. Withers, *Geography and Science in Britain 1831–1939: A Study of the British Association for the Advancement of Science* (Manchester: Manchester University Press, 2010). The figures given for the foundation of geographical societies before and after 1871 are from J. S. Keltie and H. R. Mill, *Report of the Sixth International Geographical Congress* (London: John Murray, 1895), xiii.

10. George Kish, *Bibliography of International Geographical Congresses 1871–1976* (Boston: G. K. Hall, 1979), gives the authors and titles of papers for the IGC meetings but makes no reference to meeting protocol, IGC structures, and so on.

11. Anon., "History of the Congress," in *Report of the Eighth International Geographic Congress* (Washington: Bruce and Company, 1905), 15–16.

12. The groups or sections at Antwerp were geography, cosmography, ethnography,

and the all-embracing "Navigation, Voyages, Commerce, Meteorology and Statistics." Discussion also took place across these groups. For the six meetings at Antwerp in 1871 that discussed the prime meridian, see *Compte-Rendu du Congrès des Sciences Géographiques*, 1:176, 183, 184, 206–209, 381; 2:234, 254–257.

13. [E. Ommaney], "Additional Notices," *Proceedings of the Royal Geographical Society* 16 (1871–1872): 134.

14. I have found no trace of manuscript records of the meeting and so take the printed record—inevitably a summary—as the only record of proceedings. On the discussion following Omanney's question, see *Compte-Rendu du Congrès des Sciences Géographiques*, 1:206–8 and 2:254–55.

15. *Compte-Rendu du Congrès des Sciences Géographiques*, 2:255–56.

16. G. Visconti, "Du Premier Méridien, par Otto Struve," *Bulletin de la Société de Géographie* IX (1875): 46–64; A. Germain, "Le premier meridien et la connaissance des temps," *Bulletin de la Société de Géographie* IX (1875): 504–21; Bartky, *One Time Fits All*, 43–44.

17. *Congrès International des Sciences Géographiques Tenu à Paris du Ier au 11 Août 1875: Compte Rendu des Séances*, 2 vols. (Paris: Sociéte de Géographie, 1880), 1:26–27, 29, 30; 2: 400–402.

18. Royal Geographical Society (with the Institute of British Geographers), Committee Minute Book, September 1872–October 1877, f. 136, December 17, 1874.

19. This summary and that of the preceding paragraph is based on *Congrès International des Sciences Géographiques*, 1:26–27, 29, 30; 2:400–402, and Bartky, *One Time Fits All*, 45–46.

20. A. Salomon, F. de Morsier, and L.-H. de Laharpe, "Mémoire sur la fixation d'un premier méridien," *Mémoires de la Société de Géographie de Genève* 40 (1875): 87–94.

21. On the 1876 Brussels meeting and its colonial originating contact and legacy, see Sandford Bederman, "The 1876 Brussels Geographical Conference and the Charade of European Cooperation in African Cooperation," *Terrae Incognitae* 21 (1989): 63–73. That de Beaumont was intending to give his paper on the Bering Strait prime meridian was reported upon in *Le Globe*, the journal of the Geographical Society of Geneva, at a meeting in February 1876: *Le Globe* 15 (1876): 22. The society discussed the issue the previous year, reporting that a common initial meridian and "point of [longitudinal] departure" was in the best interests of all nations: *Le Globe* 14 (1875): 216–17. The discussion of delegates to the 1879 International Congress on Commercial Geography over a single prime meridian for commercial reasons and as an advantage in maritime affairs is summarized in *Bulletin de la Société Belge de Géographie* 3 (1879): 592–96. E. Cortambert, "Selecting a First Meridian," *Popular Science Monthly* 15 (1879): 156–59, quote on 159.

22. This paragraph is based on the several papers on this issue by Henri B. Bouthillier de Beaumont: "Le méridien unique," *L'Exploration* 1 (1877): 131–32; "Choix d'un méridien initial," *L'Exploration* 7 (1879): 132–36; "Note [d'un méridien initial]," *Le Globe* 18 (1879): 202–8; and, most importantly, his 1880 pamphlet *Choix d'un Méridien Initial Unique* (Geneva: Libraire Desrogis, 1880). See also Bartky, *One Time Fits All*, 47; H. M. Smith, "Greenwich Time and the Prime Meridian," *Vistas in Astronomy* 20 (1976): 219–29. On Bouthillier de Beaumont's own discussion of the 1875

Paris Congress, see H. Bouthillier de Beaumont, "Quelques mots sur l'Exposition Géographique de Paris," *Bulletin de la Société de Géographie de Genève* 40 (1875): 210–26, esp. 210–16.

23. This meeting was, as its title suggests, a commercial and trade-based international convention with a focus on geographical apparatus, books, and atlases. This emphasis, the lack of any formal connection to the IGC meetings, and the earlier presence of the IGC in Paris in 1875 accounts for it not being seen as an IGC meeting.

24. Charles P. Daly, "Annual Address: Geographical Work of the World in 1878 & 1879," *Journal of the American Geographical Society of New York* 12 (1880): 1–107, quotes on 7–8.

25. Daly, "Annual Address," 8–9.

26. G. M. Wheeler, *Report upon the Third International Geographical Congress and Exhibition at Venice, Italy, 1881* (Washington, DC: Government Printing Office, 1883), 30–31. In his listing of the prime meridians used in topographical mapping, Wheeler has a footnote to the effect that "Spain at different epochs has counted longitude from no less than eleven distinct and separate meridians" (30). Further and different evidence of the many prime meridians then in use is to be found in different genres of topographic mapping. In the scientific cartography that informed his popular journal *Geographische Mitteilungen*, August Peterman used either the Paris or the Greenwich prime meridian before favoring Greenwich after 1885: Jan Smits, *Petermann's Maps: Carto-bibliography of the Maps in* Petermann's Geographische Mitteilungen *1885–1945* (t'Goy Houten: HES and De Graaf, 2004).

27. Sandford Fleming, *The Adoption of a Prime Meridian to be Common to All Nations: The Establishment of Standard Meridians for the Regulation of Time, Read before the International Geographical Congress at Venice, September, 1881* (London: Waterlow and Sons, 1881), quotes on 4, 6, 7, and 13, respectively.

28. The quote is from Wheeler, *Report upon the Third International Geographical Congress*, 23; see also *Società Geografica Italiana Terzo Congresso Geografico Internationale Tentuta a Venezia Dal 15–22 Settembre 1881* (Roma: Alla Sede Della Società, 1882), 1:248.

29. This is taken from the "List of Views Issued by Each Group and Not Presented at the General Sessions," in *Società Geografica Italiana Terzo Congresso Geografico Internazionale tenuto a Venezia* 1:392, and Wheeler, *Report upon the Third International Geographical Congress*, 23. See also Bartky, *One Time Fits All*, 66–67. For the first expression of A.-E. Beguyer de Chancourtois's proposal, see his "Programme d'un système de géographie," *Bulletin de la Société de Géographie* 8 (1874): 270–336. His paper to Venice in 1881 appears as "L'adoption d'un méridien initial international," *Terzo Congresso Geografico Internazionale tenuto a Venezia* 1 (1882): 20–22. See also his *Observation au Sujet de la Circulaire du Gouvernement des États Unis, concernant l'adoption d'un Méridien Initial Commun et d'une Heure Universelle* (Paris: Privately printed, 1883).

30. [T. H. Holdich] in "Geographical Notes," *Proceedings of the Royal Geographical Society and Monthly Record of Geography* 13 (1891): 615–16. On Holdich's distinguished career as an imperial surveyor, see Kenneth M. Mason and H. L. Crosthwait, "Colonel Sir Thomas Hungerford Holdich," *Geographical Journal* 75 (1930): 209–17.

31. Bartky, *One Time Fits All*, 98.

32. Norman J. W. Thrower, *Maps and Civilization: Cartography in Culture and Society* (Chicago: University of Chicago Press, 1999), 164–71; Alastair W. Pearson and Michael Heffernan, "The American Geographical Society's Map of Hispanic America: Million-Scale Mapping between the Wars," *Imago Mundi* 61 (2009): 215–43. On the public announcement of the resolution in Berne in 1891 that Greenwich should be universally adopted, see *Times*, August 14, 1891, 3.

33. D'Italo Enrico Frassi, "On Time-Reform, and a System of Hour Zones," in *Report of the Sixth International Geographical Congress: held in London, 1895* (London: John Murray, 1896), 261–68, quote on 262. Horace Bouthillier de Beaumont, "Resolution as to Standard Time," in *Report of the Sixth International Geographical Congress*, 255–56 and 259, respectively.

34. *Verhandlungen des siebenten Internationalen Geographen-Kongresses, Berlin, 1899*, 2 vols. (Berlin: W. H. Kuhl, 1901). The recommendation over Greenwich and the metric system for Penck's map project is made in 1:5. On Mill's advocacy of the metric system generally, see H. R. Mill, "On the Adoption of the Metric System of Units in All Scientific Geographical Work," in *Verhandlungen*, 2:120–24.

35. *Report of the Eighth International Geographical Congress Held in the United States, 1904* (Washington, DC: International Geographical Union, 1905), 109–10.

36. Powell, "Geographies of Science," 316; emphasis original.

37. Claire Waterton, Rebecca Ellis, and Brian Wynne, *Barcoding Nature: Shifting Cultures in an Age of Biodiversity* (London: Routledge, 2013).

38. See, for example, James A. Secord, "How Scientific Conversation Became Shop Talk," *Transactions of the Royal Historical Society* 17 (2007): 129–56; Diarmid A. Finnegan, "Placing Science in an Age of Oratory: Spaces of Scientific Speech in Mid-Victorian Edinburgh," in Livingstone and Withers, *Geographies of Nineteenth-Century Science*, 153–77.

39. For an illustration of how this is possible with respect to the meetings of the British Association for the Advancement of Science in the nineteenth century, see Withers, *Geography and Science in Britain 1831–1939*.

40. Lewis Pyenson, "An End to National Science: The Meaning and the Extension of Local Knowledge," *History of Science* 40 (2002): 251–90. On the idea of circuits and networks over and above nations, see Alan Lester, "Imperial Circuits and Networks: Geographies of the British Empire," *History Compass* 4 (2006): 124–41.

41. Peter Galison, "Ten Problems in History and Philosophy of Science," *Isis* 99 (2008): 111–24.

42. Simone Turchetti, Nestor Herran, and Soraya Boudia, "Introduction: Have We Ever Been 'Transnational'? Towards a History of Science across and beyond Borders," *British Journal for the History of Science* 45 (2012): 319–36, quote on 322. For the perspective of historians on this category, see C. A. Bayly, S. Beckert, M. Connelly, I. Hofmeyr, W. Kozol, and P. Seed, "AHR Conversation: On Transnational History," *American Historical Review* 111 (2006): 1441–64.

43. G. Kreitner, *Report of the Third International Geographical Congress, Venice [Sept. 1881]* (London: Privately published, 1883). The Austrian Gustav Kreitner attended the Venice IGC as a delegate of the North China Branch of the Royal Asiatic Society in order to receive a medal for his trans-India and Asia expedition with Bela Count Széchenyi between 1877 and 1880. Kreitner observed of the Venice meeting both that

"the almost total absence of an international character, produced, in the majority of the visitors, a marked discontent, under the weight of which the whole Congress visibly suffered," and how, in order to present the city in a good light during the Congress, Venetian municipal authorities had "kept two thousand vagabonds under lock and key": Kreitner, *Report of the Third International Geographical Congress*, 8 and 9, respectively.

44. On these examples, see Frank A. L. James, "An 'Open Clash' between Science and the Church? Wilberforce, Huxley and Hooker on Darwin at the British Association, Oxford, 1860," in *Science and Belief: From Natural Philosophy to Natural Science, 1700–1900*, ed. David M. Knight and Matthew D. Eddy (Aldershot: Ashgate, 2005), 171–93; J. R. Lucas, "Wilberforce and Huxley: A Legendary Encounter," *Historical Journal* 23 (1979): 313–30; Ruth Barton, "John Tyndall, Pantheist: A Re-reading of the Belfast Address," *Osiris* 3 (1987): 111–34; and Livingstone, *Dealing with Darwin*.

45. Secord, "Knowledge in Transit"; Livingstone, "Science, Site and Speech."

46. On these questions, see Roger Cooter and Stephen Pumfrey, "Separate Spheres and Public Places: Reflections on the History of Science Popularization and Science in Popular Culture," *History of Science* 32 (1994): 237–67.

CHAPTER SEVEN

Illustrating Nature

Exploration, Natural History, and the Travels of Charlotte Wheeler-Cuffe in Burma

NUALA C. JOHNSON

As the nineteenth century progressed, the long tradition of botanical collecting expanded, and the search for new exotic species and the desire to understand tropical nature deepened.[1] Joseph Hooker's plant-hunting expeditions to Sikkim in the northeast Himalayas, for example, a territory relatively uncharted by European explorers, aroused professional interest in new species of rhododendrons. He undertook two major trips and, on the first, was accompanied by a team of fifty-six local helpers who acted as guards, porters, cooks, and collectors. On its completion, Hooker had amassed such a huge collection of plants that it took him six weeks to catalog, arrange, and pack the eighty "coolie-loads" of material to be sent back to his father, William Hooker, director of Kew Gardens.[2] Rhododendrons formed the core of the collection, and their intense colors and bold flowers triggered a rhododendron craze in the United Kingdom. By the 1870s, they had replaced American flora in British gardens.[3]

A popular thirst to increase the supply of other exotic species meant that nurserymen as well as professional botanists were actively seeking rare plants, the former to capitalize on the commercial opportunities of such new introductions and the latter to enhance scientific understanding of tropical nature. The Veitch nursery sent out more than twenty-two plant hunters to collect specimens, and the Lobb brothers were two of the most influential collectors in the 1840s. William Lobb explored and collected in South America, while Thomas traveled to the East Indies, including Java, the Malay Peninsula, Thailand, and Burma, where he acquired new species of orchid. Similarly, George Forrest, after training in the Royal Botanic Garden Edinburgh, became a prolific explorer and plant hunter in the late nineteenth century. Over the course of his lifetime, he undertook seven expeditions to China and surrounding territories. Like Hooker, his particular specialism was collecting new rhododendrons. This reflected and drove the increased demand for the species in public, scientific, and private gardens.[4] The tradition of exploring and plant collecting continued well into the twentieth century, with Frank Kingdon-Ward, son of the Cambridge professor of botany Harry Marshall Ward, spending his lifetime hunting plants in some of the remoter parts of Assam, Tibet, and Burma, in the company of his second wife, Jean Macklin. From these excursions, he introduced new species of primula, rhododendron, cotoneaster, and lily. Kingdon-Ward famously ascended Mount Victoria in west Central Burma at the age of seventy-one.[5]

Kingdon-Ward was not the first European naturalist to scale Mount Victoria. Charlotte Wheeler-Cuffe, with her female companion Winifred MacNabb, climbed the mountain more than forty years earlier in 1911 and did so again in 1912. Wheeler-Cuffe collected many species of plant, some new and rare introductions to Britain and Ireland, and produced several landscape watercolor sketches of the expedition as well as botanical illustrations of plants amassed on the trip.[6] As she put it in a letter to her aunt in 1900: "I delight in the life out here, it is so full of interest, and the freedom and unconventionality of it suits me down to the ground."[7]

While many of the most documented nineteenth-century plant explorers were men, women played a significant, if underrecognized, role in the search for new plant species and in the visual representation of botanical specimens. Popular journals, such as *Curtis's Botanical Magazine*, employed female illustrators such as Matilda Smith, who served as the magazine's principal artist between 1887 and 1920. Women were similarly hired to illustrate for the expanding market in published *Flora*.[8]

This chapter examines the work of Wheeler-Cuffe as an "amateur" naturalist, illustrator, and explorer. It focuses, in particular, on her arrival in Burma in 1897 and the first six months of her residency there. Her initial encounters with Burmese life would shape her subsequent appetite for accumulating botanical and cultural knowledge about the region and its peoples, although she would spend twenty-five years in the colony. The chapter situates her work within the wider scholarly literature devoted to the historical geographies of scientific knowledge in the late nineteenth century and highlights the important, if often overlooked, contribution of women to the production of such knowledge.

Historical Geographies of Knowledge

In recent years, historical geographers have argued persuasively that space and place are critical to understanding the constitution, production, and circulation of science.[9] David Livingstone's work in particular has animated much of this seemingly counterintuitive claim that place is critical to the generation and consumption of scientific knowledge, and that the transmission and acceptance of this knowledge is inflected in significant ways through its geography. As Livingstone notes, "there is no single, overarching formula explaining how space invariably shapes science. . . . For in different locations, at different times, in different circumstances, and at different scales, space has made its mark on science in different ways."[10] He has pursued these claims by examining different sites of production such as the laboratory, the museum, the botanic/zoological garden, and the field. For Livingstone, these sites are also the intellectual and social milieu that shape and constitute the interpretive strategies deployed in making scientific judgments. The Royal Museum of Ethnology in Berlin, for example, was founded on the theoretical claim that cultural artifacts were of little scientific value if viewed as single items and should be seen and interpreted as part of a series. This resulted in such a vast accumulation of specimens that the site lacked any coherent order: consequently, the "very site that had cradled the developing discipline provided the structural foundation for discarding the approach to the subject it was originally established to advance."[11] Drawing on these observations, my comparative study of the botanical gardens in Belfast, Cambridge, and Dublin as locales for the development of botany demonstrated that the geographical and social context in which each was established produced different approaches to displaying and understanding natural history. It showed, too, how the debates underpinning their practices were shaped by the personnel, intellectual context, and urban setting in which each emerged.[12] To varying degrees in each place, the ambition to order nature along axioms of strict botanical classificatory systems was regularly compromised by the desire to visually represent nature along principles of aesthetic appeal, at times gesturing to the precepts of natural theology while being, at other junctures, informed by ideals of the picturesque.[13]

The "field" proved to be a foundational site for scientific investigation in the emerging Victorian sciences of geology, botany, and physical geography.[14] The attendant narratives associated with fieldwork, of manly heroics practiced often in hazardous conditions and under extended timescales, contributed to the marginalization of women from the early development of these professionalizing disciplines.[15] While this characterization of field science itself may have been the product of a longer history of European exploration and empire building in the non-European world, the practices of working in the field often defied such simple constructions as masculine endeavors. The fieldwork of Alfred Russel Wallace in the Malay Archipelago, for example, was not, as has often been claimed, an exemplar of rugged individualism but the product of entangled relationships between the preexisting colonial apparatus and the local population. Livingstone observes that "the idiom of heroic individual is all wrong. Wallace's field science was an inescapably social affair. And the knowledge he acquired was the compound product of personal observation, trusted testimony, and colonial infrastructure."[16] Wallace's own lowly position within the British social hierarchy contrasted so markedly with that of the privileged Charles Darwin that it greatly conditioned the manner in which their relative contributions as cofounders of evolutionary theory were measured and narrated, and so exemplifies Steven Shapin's assertion of the "ineradicable role of people-knowing in the making of thing-knowing."[17] In Wallace's textual representations of tropical nature, there lay, Nancy Stepan suggests, an ambivalent attitude, an "anti-tropicalism" that disrupted the conventions of narrating these regions in the languages and images of abundance, sublimity, and exoticism.[18]

While the sites of production of scientific knowledge have been investigated from a range of disciplinary perspectives, and the lexicon of geography as a subject has been embraced to inform these studies, the reception and consumption of these knowledges have also been inflected by the spatial turn in the history of science.[19] This is evident in the intimate speech spaces in which natural history societies in Victorian Scotland traded their endeavors.[20] It is clear from Livingstone's recent analysis of how religious communities of Scots Presbyterian origin engaged with Darwinism in Belfast, Edinburgh, South Carolina, and Toronto.[21] These and other studies have reinforced how the facts of place impinge upon the ways in which scientific theories are interpreted. They alert us to the significance of scale and movement in any discussion of the historical geographies of science. Site, region, nation, and globe are, according to Livingstone, at least four scales that can shape the conceptual and empirical map of how knowledge is made, and transported, across the globe.[22]

When considering the role of women in the construction and mediation of scientific knowledge, private spaces such as the home, with its associated gendered divisions of labor, the garden, and the unofficial field excursion, are all significant sites of value in understanding how women produced and translated their findings. Recent studies have emphasized how domestic spaces have operated as both sites of opportunity and constraint for women's entanglement with scientific pursuits, both as assistants to the men in their household and independently of them.[23] Women's private correspondence, and personal archives of diaries, drawings, and plant collections, often trump their public pronouncements or published treatises, and it is through these material artifacts that we can trace their contributions to natural history knowledge. In the case of Wheeler-Cuffe, the several processes of observing, recording, collecting, and painting the Burmese natural and cultural environment, and the acts of establishing venues and avenues through which she could construct and communicate that knowledge, were shaped by the uneven geographies of opportunity afforded to women practicing natural history at the turn of the twentieth century.

Space, Gender, and Natural History

In parallel with the focus on the role of space and place in the making of geographical knowledge, feminist geographers have sought to incorporate women more firmly into the canon of geography's histories.[24] They have examined the multifarious experiences and interpretations of the natural world provided by women who traveled to the colonies as missionaries, travel writers, and the wives of colonial officials. Abandoning essentializing versions of gender subjectivities, these studies have investigated how race and class, as well as gender, informed the work of prominent female travelers such as Mary Kingsley and shaped their encounters with wider imperial projects.[25] Travel, of course, was itself regularly contrasted with expeditionary science along lines that suggested "the overt masculinity of exploration and . . . [the] passive femininity of travel."[26] Affluent women such as Marianne North, Beatrix Potter, Constance Gordon-Cumming, and Theodora Guest attempted, with some success, to confront male authority by submitting their work to centers of scientific learning, such as botanical gardens and natural history societies.[27] These women might be treated as exceptional cases, however, rather than as archetypes of women's experience in engaging in scientific knowledge-making. Their familial and social status could, at times, help them defy conventional divisions of labor. Such was true of Marianne North, for example, who had monetary resources to travel extensively, to publish her findings, and to finance a gallery at Kew Gardens to display her work, or for the seventeenth-century aristocrat Mary Somerset, first Duchess of Beaufort, who collected, identified, and classified thousands of plants at her home at Badminton House.[28]

By contrast, Wheeler-Cuffe's work translated primarily through the material artifacts she collected or produced in the colonies (e.g., plant specimens, seeds, and paintings) and in the private correspondence maintained with family and friends. In this respect, she was like other relatively unrecognized female plant illustrators and collectors whose work was less visible than those who published books or essays, or who were recorded in the official documents of the colonial state. As Stoler has pointed out, the gendered nature of colonial archives has obscured the active role of women as producers of knowledge.[29] Saha agrees, claiming that female "agency has been subsumed to gendered archival and state practices."[30] Wheeler-Cuffe produced many of the plans and cross sections for her husband's engineering projects in Burma, for example, but it is his name that appears in the documents. More generally, where scientific knowledge is concerned, the valorization of printed texts and speech spaces may also obscure the contributions of women who had little opportunity to publicize their work in these ways.[31] While many official male plant hunters of the nineteenth century were supported in their expeditions to the colonies by professional institutions or commercial nurseries, women rarely enjoyed such sponsorship or imprimatur.[32] Their work was, as a result, often "invisible" in the official record and was made visible, if at all, only when they produced published accounts of their discoveries.[33]

Much existing research focuses on women travelers and on the production and reception of their work at imperial centers, as well as the celebrity status they achieved at home. This diminishes the significance of women with long-term residency in the colonies who also shaped how science was produced. In the case of Mary Kingsley, for example, Alison Blunt has suggested that Kingsley negotiated her private and public spheres differently between West Africa and Britain.[34] Kingsley's radical insights into the anthropology of West Africa contrasted with her opposition to women's suffrage and membership of the Royal Geographical Society at home in London. This has led Avril Maddrell to conclude that "her writing is riddled throughout with contradictory relationships between gender and colonialism."[35] Conversely, Laura Ciolkowski suggests that such gender ambiguities were carefully cultivated: they help to render Kingsley the model English woman, whereas the female traveler of Kingsley's text is a modern Britannia who, as "invincible global civilizing agent," symbolically links "the seemingly incompatible values of masculine strength and feminine moral leadership in the service of British imperial power."[36] In the case of Wheeler-Cuffe, I seek to reinstate as significant the consideration of women who vigorously pursued their scientific interests in the field, and, of her, in one of the remoter corners of the British Empire, yet who operated independently of formal structures of knowledge-making and who had more limited public recognition during their lifetimes.

Wheeler-Cuffe's encounters with Burma also intersect with the wider scholarly literature on the construction of the tropics as a discursive field, a material space, and a hermeneutic endeavor.[37] Historians have delineated the various ways in which the tropics have been conceived over time, from an earthly paradise of environmental luxuriance to a pathological space of disease, lethargy, and moral laxity.[38] Geographers have stressed the ways in which the idea of the tropics was conceived differently, if not also differentially, across the globe, from Alexander von Humboldt and Charles Darwin in the nineteenth century, to Pierre Gourou in the twentieth century.[39] Most of this work focuses on the experience and representation of the tropics by male naturalists, surveyors, travel writers, artists, and colonial administrators. While these studies offer a rich understanding of how men interpreted the tropics, it is by no means self-evident that women mirrored these experiences.[40] By examining Wheeler-Cuffe's reaction to the tropics, we can begin to unravel the role of gender in fashioning Western conceptions of the tropics. The practice of being a plant collector, explorer, gardener, and artist, alongside the quotidian duties of the wife of a colonial official, marks Wheeler-Cuffe as different from professional plant collectors (female or male) sent from Britain to seek out botanical treasures. The focus on material, textual, and visual artifacts, gathered and produced in situ, provides a different frame for understanding her knowledge-making activity than the work of travel writers whose texts, often, were produced, edited, and distributed at home for domestic audiences.

Wheeler-Cuffe's position enabled her to straddle the spaces of public consort and private resident, interacting with the officialdom of empire but also developing strong links with the indigenous population (servants, pupils in schools where she taught painting, guides, Burmese officials), upon whom she relied, in various ways, to explore the colony's natural history. Her experience in Burma deviates from traditional historiographies that claim that "science and colonization can be precisely expressed through the term *colo-*

nial science."[41] It reinforces a growing body of research that focuses on the complex network of entanglements between metropole and colony.[42] How science travels across space in the nineteenth century is, as historical geographers have proposed, neither unidirectional nor uniform.[43] The natural history produced by Wheeler-Cuffe, and her position within the wider world, reflects Tony Ballantyne's concept of "webs of empire."[44] Her practice of science in the colonies, I suggest, was in conversation with, rather than determined by, the science of the metropole. Through a circuit of social and intellectual connections, Wheeler-Cuffe produced knowledge about Burma's natural history that was as much a result of her role as the wife of a colonial official, her contact with the indigenous population, and her ancestral and professional links with naturalists in Ireland and Britain as of any direct guidance provided by the imperial center.[45]

Locating Charlotte Wheeler-Cuffe

From specimen collections and travel diaries to illustrated drawings and handbooks of flora and fauna, women were the creators of a significant corpus of knowledge about the natural world, both at the local scale of the domestic sphere and on a more global scale through long-distance travel or being domiciled overseas as part of Britain's colonial ventures.[46] Although there was an increasing differentiation between the amateur naturalist and the professional scientific botanist in Britain after 1870, vestiges of the earlier undifferentiated category of "natural historian" lingered well into the twentieth century.[47] For women in particular, this tradition of generating knowledge through polite botany remained important as their access to the professionalizing centers of accumulation and calculation of botanical and zoological research continued to be limited.[48] As Livingstone notes, "the amateur/professional polarity could itself operate as a means of excluding women from serious scientific visibility and underscore the presumption that, for women, natural history was nothing more than a genteel hobby."[49]

Botany was considered the feminine science par excellence and was most typically the area of scientific investigation in which women could engage, as it cultivated an interest in the outdoors, disciplined the body and mind, and stimulated a culture of self-improvement.[50] Detailed studies of Victorian women naturalists also indicate that, from time to time, they stepped outside these conventional tropes of "natural" feminine pursuits in ways that challenged gendered scientific authority and the highly gendered practices associated with the observation and documentation of the natural world.[51]

Isla Forsyth has suggested that "the *where* of scientific practice at times was liberating, the colonies in particular affording women space for practicing science."[52] For Wheeler-Cuffe, her extended residency in Burma certainly provided opportunities for exploration and travel that she would not have enjoyed at home in Britain or Ireland, and she recognized this at the time. The remainder of this chapter focuses on her first six months of residency in the colony. Although much of her artistic output dates from later years, this formative initial period provides insights into how she developed her network of contacts, responded to the new tropical environment, and set the foundations for her collecting and painting career.

Wheeler-Cuffe, known affectionately as "Shadow," was elected fellow of the Royal Geographical Society in April 1922 at a time when women still represented only 7.7 percent of the society's membership.[53] Her election was in recognition of her contribution to collecting, botanical illustration, and the geographical and anthropological knowledge she accumulated about Burma during the quarter of a century (1897–1922) she spent there. Maddrell suggests that "those who travelled with husbands or family on imperial duty . . . represent a continuation of the strong link between the RGS and state institutions and interests."[54] So, what of Wheeler-Cuffe's background?

She was born in Wimbledon, Surrey, May 24, 1867, the youngest daughter of William Williams, a solicitor and one-time president of the Law Society. Her mother, the second wife of Williams, was Rose Isabella ("Rosebelle") and was the third daughter of Reverend Sir Hercules Richard Langrishe (1782–1862), third Baronet of Knocktopher, County Kilkenny. It was this Irish connection on her mother's side that, in childhood, brought her regularly to Ireland and would stimulate a long-standing relationship with and, eventually permanent residency on, the island. Charlotte and her sister Rosabelle Mary (known as "Monie") and her half sisters Alice, Amy, and Rosemary were brought up in the family home, Parkside House, in Wimbledon. They also spent time at their country house, Upperfold, in West Sussex, which their father had bought in 1889. While there is little extant information on the details of their childhood and their education, Charlotte and Monie spent much time visiting Ireland. In the 1880s, they stayed at Fir Grove near Thomastown, where their aunt, Charlotte Langrishe (known as "Cha"), lived, and at Woodstock in Inistioge, both in County Kilkenny. These ancestral connections position her in a different political space from either her English or her Scottish contemporaries. As the Home Rule movement gathered pace at the end of the nineteenth century, Irish Protestants often faced conflicting loyalties.

Wheeler-Cuffe reflected these ambivalences, and even though her political leanings were Unionist, she chose to retire to Ireland, where she died in 1967.

During her youth when visiting Ireland, Wheeler-Cuffe produced some of her earliest watercolors. On a trip to Connemara, she completed a painting of the Bundara River titled "Among the Hills, the Bundara River, Connemara" in 1888 (fig. 7.1). In it, she depicted the rapidly cascading river, Abhainn Bhun Dorcha, flowing from south Mayo into Killary Harbour, with the Mweelrea mountains and Sheeffry Hills in the background. The painting is an indicator of her developing artistic skills. It contains on the reverse a commentary on the quality of the composition by William Frank Calderon. Calderon was the son of the well-known painter Philip Hermogenes Calderon, and he ran an art school on Baker Street, London. He wrote, "A vigorous sketch. . . . The Rocks are too equally purple—in nature you would find much more variety of colour in them. . . . The distance is very well done."[55] This suggests that Wheeler-Cuffe attended art classes run by Calderon and is the only evidence that she received any formal training.

Aged thirty, she married Otway Wheeler-Cuffe, whom she had known since childhood. When the marriage was imminent, she wrote to her Aunt Cha about the prospect of moving to Burma, where Otway was already stationed: "If we went out this year, we could come home next, when his furlough will be due. . . . It would be sort of a trial trip to see how the climate agreed with me & if it really did <u>not</u> then it would be perhaps possible to make some other arrangement—Though I believe a hot climate will rather suit me than otherwise."[56] From its outset, a life in the colonial service was a prospect she relished, filling her with enthusiasm rather than anxiety. While climate was the defining characteristic in what Livingstone calls the "anticipative geography of the tropics,"[57] we will see that Wheeler-Cuffe's encounter with the weather of Burma proved most agreeable.

Otway Wheeler-Cuffe was born on December 9, 1866, in Southsea, Hampshire, but likewise had close connections with Ireland. His parents, Major Otway and Mrs. Cuffe, lived at Woodlands, a house near Waterford, and his uncle, Sir Charles Wheeler-Cuffe, was the second baronet of Leyrath, an estate located a few miles from Kilkenny city. This baronetcy, established in 1800, was the last ever created in the Baronetage of Ireland, and, on the death of his uncle in 1915, he inherited the title and land at Leyrath.[58] The title became extinct upon his death in 1934, but Charlotte continued to live on the estate until her death. In 1886, Otway Wheeler-Cuffe enrolled in the Royal Indian Engineering College, Coopers Hill, where he trained as a civil engi-

Figure 7.1. Charlotte Wheeler-Cuffe, "Among the Hills, the Bundara River, Connemara" (1888). Courtesy of the National Botanic Gardens, Glasnevin, Dublin.

neer. After qualifying, he spent a year working on a ship canal before entering the Public Works Department in 1890 and being stationed in Burma. He and Charlotte married on June 3, 1897, at Lodsworth parish church near her family's country house, Upperfold. They set sail ten days later from Liverpool to Rangoon on the Bibby Line's SS *Staffordshire*.[59] Her life in Burma was about to begin. James Duncan observed that few British women settled in the highlands of Ceylon, where tropical nature was regarded as particularly hazardous, until the 1880s.[60] Similarly, in Burma it was only in the last decade of the nineteenth century that women began to arrive in numbers to the colony.

Arriving in Burma

The route of their first voyage to Burma was the one they would continue to take on their trips home over the next twenty-five years. From Liverpool, the ship traveled to the Mediterranean and stopped at Marseilles before proceeding, via the Strait of Bonifacio, into the Tyrrhenian Sea and through the Strait of Messina, where Charlotte reported getting a good view of Mount Etna. At Port Said, she commented on the multicultural nature of the place, where one witnessed "every race under the sun & every language all going at once."[61] Having loaded coal and cargo, the ship moved through the Suez Canal and into the Red Sea, past the Maldive Islands before arriving at Colombo.[62] From Colombo, the ship took the four-day journey across the Bay of Bengal until it arrived at the mouth of the Rangoon River in Burma. The total journey lasted a month.

Wheeler-Cuffe undertook this voyage many times and became accustomed to a routine of exercise, socializing with other passengers also in colonial service, taking care of her cats and dogs, and, above all, reading, writing, and sketching. On her first trip, she received from the ship's captain a copy of George W. Bird's recently published *Wanderings in Burma* (1897). Bird had served in the Educational Department for twenty years and was an acquaintance of Otway.[63] His book, which included many maps and other illustrations, provided Charlotte with an early introduction to the colony and information on the railway routes and the main tourist sites.[64] She would often produce maps to aid her husband's engineering projects and included sketch maps in her correspondence home. She illustrated the Bay of Bengal and its principal coastal features, settlements, and islands (fig. 7.2). This reflected the much-longer naval tradition of illustrating coastal features during voyages of exploration, such as those undertaken by the professional artist Augustus Earle in the second *Beagle* trip to South America.[65]

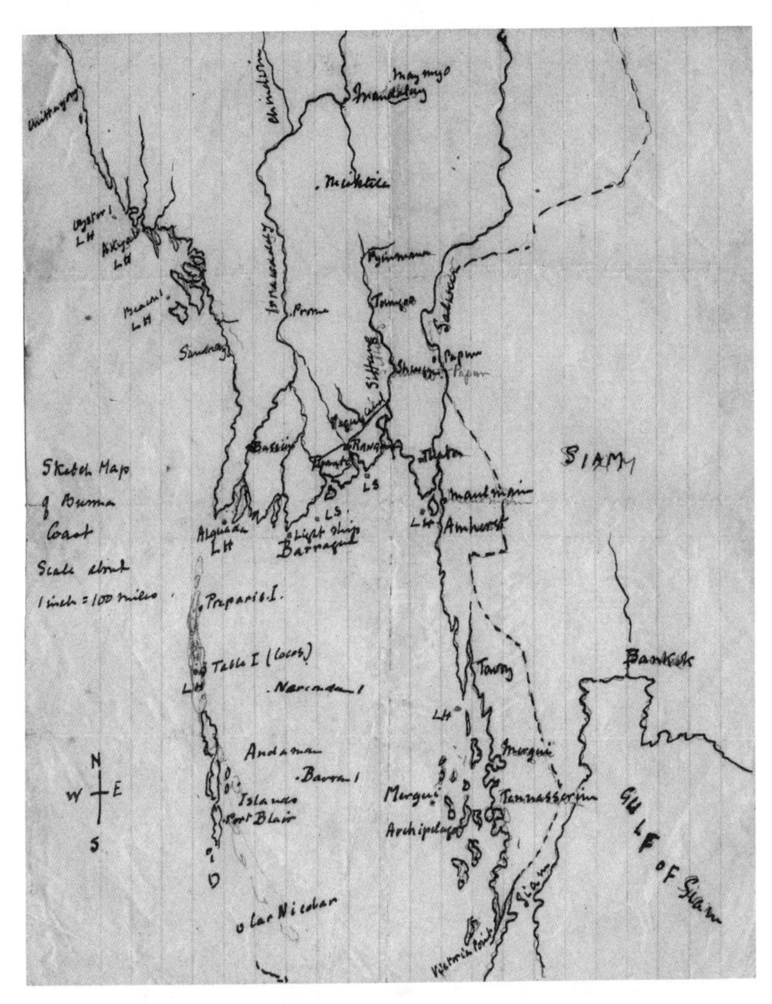

Figure 7.2. Charlotte Wheeler-Cuffe, Sketch of Burma's coast and islands. Courtesy of the National Botanic Gardens, Glasnevin, Dublin.

On arrival in Burma, the Wheeler-Cuffes were immediately stationed in Thayetmyo, a town located about two hundred miles north of Rangoon on the east bank of the Irrawaddy River. During the final decades of the nineteenth century, the Irrawaddy Flotilla Company made Burma's river easily navigable, and she arrived at her destination by steamboat. She was beguiled by the place immediately. An early watercolor captures the physical allure of the great expanse of the river as well as the local boats and boatmen working on it (fig. 7.3). In contrast to earlier Humboldtian views of the tropics, which were "essentially vegetative" and usually devoid of local people, Wheeler-

Figure 7.3. Charlotte Wheeler-Cuffe, “From the flood gauge, Thayetmyo” (1900). Courtesy of the National Botanic Gardens, Glasnevin, Dublin.

Cuffe immediately illustrated the everyday life of Burmans toiling on the great waterway.[66] She commented that "the place is charmingly pretty, & this house, though not actually on the river bank, is not very far inland."[67] In contrast to those who viewed the tropics as spaces of pestilence and languor, Wheeler-Cuffe found energy and beauty.[68] She noted the large number of servants at her disposal, and as a newcomer, she immediately set about getting acquainted with the other colonial settlers. The deputy commissioner, Mr. Carter, whom she noted was a great gardener, had already promised to provide her with plants for her "compound." She explained that her garden sloped away from the house and had "3 tall 'toddy' palms, several neem trees, something like ash, & others which I don't know. There are some fine orchids on the trees, brought in from the jungle."[69] She included in her letter a sketch of their house, which, she observed, was constructed of wood, apart from the ground floor, which was concrete. Clearly, organizing the domestic space, for which she would have had principal responsibility, was her initial priority upon her arrival. To her Aunt Cha, she provided the following description: "The place is extremely pretty and there are dozens of lovely bits to paint, though only a few that are particularly characteristic of the country and some that might be at home until you look closely at the trees & see that they are different. The place consists of a picturesque native town . . . and the rest is just like a big and very pretty park, with barracks, church & houses dotted about, each in its own 'compound,' or bit of garden."[70]

Settling into Life in Thayetmyo

From the outset, it is clear that Wheeler-Cuffe was enamored by her new abode. Determined to deepen her interest in gardening and botany, she was inspired by her surroundings to understand and illustrate the natural history and cultural landscape of the region. Her husband's role as district engineer facilitated her botanical interests. He regularly traveled around the district, which extended approximately thirty-five miles in each direction from Thayetmyo. He principally inspected and oversaw the building, management, and maintenance of roads, bridges, and official buildings deemed necessary for the successful administration of the empire in Burma. She accompanied him on these excursions, and they would provide her with the opportunity to develop a detailed knowledge of the land, its botany, and its peoples. On occasion, she would travel without Otway, in the company of other colonial wives, as she did when making her ascent of Mount Victoria in 1911.[71]

Even though she was occupied thus with establishing her domestic space, Wheeler-Cuffe undertook her first excursion with her husband within a week of their arrival at Thayetmyo. In this respect, she differed from many other colonial women who devoted themselves to establishing British domesticity overseas.[72] Wheeler-Cuffe and her husband crossed to the eastern shore of the Irrawaddy River on a "country boat," where they were met by the Public Works Department subdivisional officer Bapu Gupta. Charlotte observed that she had never felt so well in her life and developed a large appetite due to the heat.[73] Her reaction to the climate was the inverse of those Europeans who mobilized a discursive performance of the tropics as a space of indolence and disease. The party consisted of a bullock cart that transported baggage and equipment, as well as a cook, syce (groom), and Public Works Department chaprassi.[74] She rode a gray pony to Allanmyo, two miles north, where she inspected the bridges, roads, and culverts. The following day, Otway and Bapu Gupta traveled ten miles south from Yuatoung, while Charlotte remained in Allanmyo and began sketching. On the third day of their excursion, Otway and Charlotte rode to Kyaukpadoung, fourteen miles east of Yuatoung, on the road toward Toungoo. While the men inspected a guardhouse, Charlotte botanized. She improvised using a sliced and pointed bamboo as she walked along the road trying to "dig up ferns, but found they wanted a more substantial weapon."[75] Otway diverted some coolies from road remetaling to aid in her collecting. She recorded that she "came back in triumph with great spoils—a lot of climbing ferns, a lovely big thing with a flower like an orchid, something like a lily, & something like Solomon's seal in growth."[76]

Her first excursion into the field in Burma clearly and immediately presented her with unfamiliar species. The experience only whetted her appetite for undertaking further investigation and botanical collecting. Her reaction reinforces the suggestion that tropical nature, for some, was a source of intoxication and wonderment.[77] Despite being newly arrived, Wheeler-Cuffe did not confine herself to the domestic sphere and the conventional responsibilities of colonial homemaking but instead began, as she intended to continue, to combine the gendered duties of the domestic with opportunities for exploration. The most exciting discovery she reported was "the thing of all others that I haven't got the root of yet, as it was too far but I got the blossom yesterday, and that is 'Gloriosa'. . . . It was, growing by the roadside amongst a tuft of the big white thing."[78] A native to tropical and southern Africa, as well as temperate and tropical Asia, the *Gloriosa superba* was a climber with spectacular red-and-yellow flowers and was used in traditional medicine on both continents, although all parts of the plant, especially its tubers, are extremely poisonous.[79] For Wheeler-Cuffe, seeing the plant in the wild for the first time was a pleasure and contrasted with her knowledge of the species as a cultivated, ornamental exotic.

To transport the plants back to Thayetmyo, she collected a pile of old flowerpots and recruited local labor, including the chaprassi "and half a dozen other people."[80] She noted that the Burmans "all seem fond of flowers and were quite pleased at my wanting to take plants home, and most anxious to help."[81] While the social hierarchies between native and colonizer were being maintained, she was appreciative nonetheless of the cooperation and knowledge of the local people. They informed her that there were numerous ground orchids in the area, which they would collect for her the next time they visited. She was very enthusiastic at the prospect of collecting more plants, and she quickly recognized that planting her garden in Burma would be radically different from gardening at home. She observed that "most people in Thayetmyo go in for trying to grow home flowers, but they don't really do well, so I think I shall try collecting things on our various expeditions instead, and only grow home things that really flourish and not break my heart over things that die."[82] While European settlers sought to emulate the gardens of their homelands in the tropics,[83] Wheeler-Cuffe was keen to avoid mimicking the gardening aesthetic she had brought from home and embraced instead the challenge of learning about Burma's natural history and cultivating indigenous species. Her garden would be the space in which she could trial and observe the growing habits of endemic species, and, as such, would become a site for knowledge production. While Duncan claims, of the hill station in Nuwara Eliya in the Highlands of Ceylon, that "the British could create a . . . mimetic portrait of Britain," Wheeler-Cuffe was anxious to embrace the cultures of cultivation that accommodated indigenous species and exotic planting regimes.[84]

After her first excursion, Wheeler-Cuffe began a pattern of circulating plants she had found between the colonies and home. She sent her mother some seeds of a species she described to be "like mallow." On the domestic front, she noted she did not have an ayah (nanny), as the sweeper's wife carried out chores such as cleaning her bathroom and washing her clothing. She pointed out that "[as] we are out in the jungle half of our time, an ayah would be very little use, for one could not take her as she would only add extra expense. I prefer having an extra pony!!"[85] Clearly, for Wheeler-Cuffe a life of domesticity was not what she envisaged in the colonies. From the outset, she established a routine of exploration that would continue throughout her life in Burma. Maintaining contact with the affairs of the homeland was important though, and English newspapers were readily available in Thayetmyo, which, she observed, might not have been the case in a less "civilized place."[86] Although captivated by Burma, she was also inclined, like her contemporaries, to make sense of the colony by situating its regions along a continuum between tradition and modernity.

Descriptions and illustrations of Burma in picturesque idiom had gathered pace as increasing numbers of travel writers visited by the turn of the century. As Stephen L. Keck claims, travel writers "relied upon the picturesque to make Burma's tropical landscapes palpable [but] these writings indicate that the picturesque itself was a mutable concept."[87] Wheeler-Cuffe also combined the trope of the picturesque in her descriptions of the landscape, but she also provided a descriptive narrative style that was more factually based and documentary in focus. She and her husband, for instance, undertook a sixteen-mile ride to inspect the remetaling of the Mindon Road. While the waterways and their steamers enabled the British to establish a foothold in Burma, the railways were regarded as foundational to its economic exploitation, and the roads were critical for the transport of goods and people around these other transport networks. Wheeler-Cuffe fully appreciated the significance of infrastructure to maintaining the colony but also noted that poor roads were "terribly hard on the country folk and their bullock carts who have to get on as best they can on the sides of the roads."[88] As a keen animal lover, she was impressed by how the Burmese seemed "wonderfully kind" to their animals and noted also how domestic animals looked upon humans as "friends and not enemies."[89] This reciprocal respect between human and beast that she observed in Burmese society corresponded with her own attitudes toward the animal world. Despite her husband being a keen big-game hunter, she maintained a menagerie of domestic animals, including four ponies, four dogs, and two cats. The fact that the indigenous population seemed to share a respect for animals endeared her to the local population.

Her first September in the colony would prove very warm, but, rather than complaining about the tropical climate, Wheeler-Cuffe quickly adapted to it. She observed that "one has to do things in a more leisurely way in this heat and I find I can't skip into my clothes in the same hurry as at home. . . .

Still I don't really mind it that much."[90] The extraordinary shapes and colors of the landscape, produced by the weather conditions, compensated for any other inconveniences. One evening she witnessed a "beautiful [storm] on the other side of the river. . . . The great boiling masses of clouds were still tinged with crimson from the sunset while through and around them the lightening darted like fiery snakes."[91] Such tropical storms beguiled Burma's new resident and deepened her appreciation of its landscape. Otway confirmed her rapid adjustment to tropical life: "[she] delights in jungle life and is the best wife a Public Works Dept Officer could wish for."[92] For many women, newly arrived in the colonies and accustomed as they were to the regulated domesticity of home, adapting to the climate could be a challenge. Otway observed that she "delights in [the heat], unlike most people who growl all day and all night," and her "sketchbook [w]as growing daily."[93] Clearly, she readily adjusted to her new environment and took advantage of the opportunity to pursue interests outside of the domestic arena by regularly accompanying her husband on his duties and recording her observations in text, sketch, and artifact. She fashioned the space of the colonial wife in ways that differed from the conventional behavior of women at home, and, in contrast to other colonial wives, she did not attempt to reproduce that pattern overseas.

As time passed, she deepened connections with the expatriate community, especially those who shared her interests, for example, the chaplain and his wife (the Corys), who were "keen on science in general and natural history in particular."[94] They had spent six years stationed in Madagascar and "have lovely collections of the birds and butterflies they made themselves," and she had been "out sketching a couple of times" with Mrs. Cory.[95] She also began to ingratiate herself with the rest of the community by becoming the polo stick mender while Otway enthusiastically participated in the sport. Such roles helped her to become part of the wider circle while at the same time affording her the space to maintain her personal interests in natural history, plant hunting, and painting.

In addition to interacting with the English population, Wheeler-Cuffe was "beginning to get along very well with the Burmese and can understand a great deal—quite enough to ask my way in the jungle."[96] Unlike the transient travel-writer or professional plant hunter, who relied on interpreters to communicate with the indigenous population, she would become fluent in the language and, additionally, learnt Hindustani. She appreciated the value of learning the native language and, in later years, would produce a beautifully illustrated calendar of the Burmese alphabet. As their first Christmas in Burma approached, Charlotte determined that she would participate in the community's efforts to celebrate the occasion. Unlike the gendered image of scientific botany as a masculine, often solitary, enterprise, Wheeler-Cuffe understood that she could not pursue such a life path in Burma and so developed a strategy that would partly conform to the gendered practices of settlement in the colony while, at the same time, exercising her ambition to explore and record its flora. During her first Christmas, she acknowledged that she "like[d] the life out here very much, though just this week there has been rather more going on than I feel at all inclined for. But in a small community like this one cannot hold aloof without being selfish."[97] She struck a balance between interacting with the other colonial wives and the domestic sphere that often preoccupied them, and pursuing her love of botanizing in the remoter regions of the colony. To that end, she informed her father that she and Otway were undertaking a trip after Christmas to a "very out of the way place."[98] This was north of Mindon, close to the Arakan Yoma Mountains, where "the scenery is very beautiful . . . but there are no roads, so one can only go there in the dry weather when the jungle paths are passable."[99]

To some degree, her appetite for adventure and for accompanying her husband on his duties marked her out as deviating from the more orthodox gender roles of the settlement. She was not going to be confined to the compound and to a life of colonial domesticity. While she participated in communal activities and observed the niceties governing social interactions, she also created the space to pursue her own botanical interests. She sketched unabashedly in these early months and requested a new sketchbook be sent from "Bowdens (between Harveys and Harrods) . . . where I get them always and they are in a solid block, bound so that when separated they remain in the book. They are covered in brown Holland, with a place for a pencil. . . . It ought to be packed in tin foil or oiled paper, to prevent getting mildewed on the journey."[100] While the colonies provided inspiration for her art, the materials required to fulfil her passion had to come from home, and over the next twenty-five years regular shipments of art materials traveled across the Indian Ocean to Burma. During these first six months, she spent her time getting to know the place as well as establishing her domestic arrangements and planting a garden. The period marked the beginning of her efforts to record the landscape and its botanical treasures. Her detailed plant illustrations would emerge as she became better acquainted with the geography of Burma and its flora (fig. 7.4).[101]

Figure 7.4. Charlotte Wheeler-Cuffe, *Acriopsis javanica,* Karen Hills (1902). Courtesy of the National Botanic Gardens, Glasnevin, Dublin.

Conclusion

In recent years, historical geographers interested in the where and the how of geographical knowledge production have produced a wealth of insights into how various types of scientific and geographical theories and practices have been produced, contested, and circulated differentially across the globe. Livingstone's magisterial analysis of the significance of place, politics, and rhetoric in shaping the response to Darwinian theories marks the culmination of his long engagement with understanding the relationships between geography and science.[102] Historical geographers' analyses of how knowledge was disciplined through regimes of observation, measurement, and instrumentation have highlighted how authoritative knowledge was produced and circulated globally. At the same time, however, emphasis on the published text and image, or on the performance of scientific discourse and practice through speech spaces and public demonstrations of natural phenomena, has contributed to the marginalization of certain voices in the histories of knowledge-making. The role of indigenous expertise in the construction of scientific theories and practices, and the significance of women as knowledge makers, while receiving some critical attention recently, are both largely secondary in much empirical research. In part, this is a consequence of the gendered practices of the scientific communities under investigation and the resulting absences in the archival record, especially in governmental sources. Uncovering the "hidden" labors and laborers requires a focus on different modes of engagement and reading "against the archival grain."[103] Feminist geographers and others have gone some way toward highlighting the types of knowledge produced by nineteenth-century Western female travel writers and naturalists who spent time traversing the globe.

We can draw the following conclusions of Wheeler-Cuffe's initial encounter with Burma. First, for Wheeler-Cuffe the tropics did not represent a pathological space to be endured. On the contrary, she relished the heat and humidity of her new environment and quickly adjusted to its routines and conventions. The tropical landscape, with its rich botanical treasures, was more or less instantly a source of inspiration for her artistic work and collecting practices. Unlike plant hunters who traveled to South Asia to collect solely for commercial ends, Wheeler-Cuffe's activities were not guided by such narrowly defined objectives. Her long-term residency in Burma allowed her to appreciate tropical nature and its biogeography on its own terms. Consequently, her knowledge base was firmer and wider than that of the more transient plant hunter, and her connections to the colony deeper rooted.

Second, as a colonial resident stationed in many different regions of Burma, Charlotte Wheeler-Cuffe developed close linkages with the indigenous population. These connections transcended the more superficial ones established by plant hunters who hired locals to facilitate their expeditions. She became fluent in Burmese. She instructed local children in painting and drawing. She developed a botanical garden near Maymyo in 1918. She became well acquainted with Burmese cultural and religious practices. Although she remained a loyal servant of empire throughout her residency, her knowledge of the native cultures, developed over a long time, modulated her attitudes toward the virtues of imperialism and its practices in this corner of South Asia.

Finally, colonial science differed from the practice of science at the imperial center as it provided opportunities that liberated Wheeler-Cuffe from the strict gender constraints of late Victorian Britain. She frequently joined her husband on his daily routines as district engineer and used these trips to botanize and to make observations about the region's natural history. The prospect of her accompanying him to work had they remained at home is remote. The colonies enabled her to behave in ways that would not have found favor in Britain, and it was through this persistent fieldwork that she amassed much of her knowledge. More broadly, colonial space provided her and other women, including some of the wives of other administrators in the colonies, with the potential for disrupting the gender norms prevailing in Britain. Being at some distance removed from the social and political center emancipated her, to some degree, from the stricter regulatory regimes of gendered conformity policed and practiced at home.

Notes

1. David Arnold, *The Tropics and the Travelling Gaze: India, Landscape, and Science, 1800–1856* (Seattle: University of Washington Press, 2006); Nancy L. Stepan, *Picturing Tropical Nature* (London: Reaktion, 2001); Felix Driver and Luciana Martins, eds., *Tropical Visions in an Age of Empire* (Chicago: University of Chicago Press, 2005).
2. Jim Endersby, *Imperial Nature: Joseph Hooker and the Practices of Victorian Science* (Chicago: University of Chicago Press, 2008).
3. Toby Musgrave, Chris Gardner, and Will Musgrave, *The Plant Hunters: Two Hundred Years of Adventure and Discovery around the World* (London: Lock, 1998).
4. Charles Lyte, *The Plant Hunters* (London: Orbis, 1983); John McQueen Cowan, ed., *The Journeys and Plant Introductions of George Forrest* (Oxford: Oxford University Press, 1952).
5. Charles Lyte, *Frank Kingdon-Ward: The Last of the Great Plant Hunters* (London: John Murray, 1989).
6. Nuala C. Johnson, "On the Colonial Frontier: Gender, Exploration and Plant-Hunting on Mount Victoria in Early 20th-Century Burma," *Transactions of the Institute of British Geographers* 43 (2017): 417–31.
7. Letter from Wheeler-Cuffe to Aunt Cha, August 12, 1900, Thayetmyo, National Botanic Gardens Dublin, Archive (hereafter NBG).
8. Wilfred Blunt and William T. Stearn, *The Art of Botanical Illustration* (London: Collins Antique Collectors Club, 1993); J. M. P. Brenan, *A Vision of Eden: The Life and*

Work of Marianne North (New York: Holt, Rinehart and Winston, 1980); Ann B. Shteir, *Cultivating Women, Cultivating Science: Flora's Daughters and Botany in England, 1760–1860* (Baltimore: Johns Hopkins University Press, 1996).

9. David N. Livingstone, *Putting Science in Its Place* (Chicago: Chicago University Press, 2003); Diarmid A. Finnegan, *Natural History Societies and Civic Culture in Victorian Scotland* (London: Pickering and Chatto, 2009); Charles W. J. Withers, *Placing the Enlightenment: Thinking Geographically about the Age of Reason* (Chicago: University of Chicago Press, 2007).

10. Livingstone, *Putting Science in Its Place*, 4.

11. Livingstone, *Putting Science in Its Place*, 37.

12. Nuala C. Johnson, *Nature Displaced, Nature Displayed: Order and Beauty in Botanical Gardens* (London: I. B. Tauris, 2011).

13. Nuala C. Johnson, "Grand Design(er)s: David Moore, Natural Theology and the Royal Botanic Gardens in Glasnevin, Dublin, 1838–1879," *Cultural Geographies* 14 (2007): 29–55.

14. On which point, see Jan Golinski, *Making Natural Knowledge: Constructivism and the History of Science* (Cambridge: Cambridge University Press, 1998), and the essays in Kristian H. Nielsen, Michael Harbsmeier, and Christopher J. Ries, eds., *Scientists and Scholars in the Field: Studies in the History of Fieldwork and Expeditions* (Aarhus: Aarhus University Press, 2012). On the early development of disciplines/subjects in the nineteenth-century field sciences, see David Cahan, ed., *From Natural Philosophy to the Sciences: Writing the History of Nineteenth-Century Science* (Chicago: University of Chicago Press, 2003).

15. Avril Maddrell, *Complex Locations: Women's Geographical Work in the UK 1850–1970* (Oxford: Wiley-Blackwell, 2009); Mark Carey, M. Jackson, Allesandro Antonello, and Jaclyn Rushing, "Glaciers, Gender and Science: A Feminist Glaciology Framework for Global Environmental Change Research," *Progress in Human Geography* 40 (2016): 770–93.

16. Livingstone, *Putting Science in Its Place*, 44.

17. Quoted in David N. Livingstone, "Text, Talk and Testimony: Geography Reflections on Scientific Habits: An Afterword," *British Journal for the History of Science* 38 (2005): 97.

18. Stepan, *Picturing Tropical Nature*, 57–84.

19. Diarmid A. Finnegan, "The Spatial Turn: Geographical Approaches in the History of Science," *Journal of the History of Biology* 41 (2008): 369–88.

20. Diarmid A. Finnegan, "Placing Science in an Age of Oratory: Spaces of Scientific Speech in Mid-Victorian Edinburgh," in *Geographies of Nineteenth-Century Science*, ed. David N. Livingstone and Charles W. J. Withers (Chicago: University of Chicago Press, 2011), 153–77.

21. David N. Livingstone, *Dealing with Darwin* (Baltimore: Johns Hopkins University Press, 2014).

22. Livingstone, *Putting Science in Its Place*, 1–16.

23. Donald L. Opitz, Staffan Bergwik, and Brigitte Van Tiggelen, eds., *Domesticity in the Making of Modern Science* (Basingstoke: Palgrave Macmillan, 2016); Donald L. Opitz, "Domestic Space," in *A Companion to the History of Science*, ed. Bernard Lightman (Oxford: Wiley Blackwell, 2016), 252–67.

24. Mona Domosh, "Towards a Feminist Historiography of Geography," *Transactions of the Institute of British Geographers*, 16 (1991): 95–104; Gillian Rose, *Feminism and Geography: The Limits of Geographical Knowledge* (Minneapolis: University of Minnesota Press, 1993).

25. Alison Blunt, *Travel, Gender and Imperialism: Mary Kingsley and West Africa* (London: Guilford Press, 1994); Cheryl McEwan, *Gender, Geography and Empire: Victorian Women Travellers in Africa* (Farnham: Ashgate, 2000).

26. Blunt, *Travel, Gender and Imperialism*, 32.

27. Jeanne K. Guelke and Karen M. Morin, "Gender, Nature, Empire: Women Naturalists in Nineteenth-Century British Travel Literature," *Transactions of the Institute of British Geographers* 26 (2001): 306–26; Karen Morin, "Peak Practices: Englishwomen's 'Heroic' Adventures in the Nineteenth-Century American West," *Annals of the Association of American Geographers* 89 (1999): 489–514; Linda Lear, *Beatrix Potter: A Life in Nature* (London: Penguin, 2006); Suzanne Le-May Sheffield, *Revealing New Worlds: Three Victorian Women Naturalists* (London: Routledge, 2001).

28. Julie Davies, "Botanizing at Badminton House: The Botanical Pursuits of Mary Somerset, First Duchess of Beaufort," in *Domesticity in the Making of Modern Science*, ed. Opitz, Bergwik and Tiggelen (Basingstoke: Palgrave Macmillan, 2016), 19–40.

29. Ann Stoler, *Along the Archival Grain: Epistemic Anxieties and Colonial Common Sense* (Princeton, NJ: Princeton University Press, 2010).

30. Jonathan Saha, "The Male State: Colonialism, Corruption and Rape Investigations in the Irrawaddy delta c. 1900," *Indian Economic and Social History Review* 47 (2010): 345.

31. Roger Cooter and Stephen Pumfrey, "Separate Spheres and Public Places: Reflections on the History of Science Popularization and Science in Popular Culture," *History of Science* 32 (1994): 237–67.

32. Musgrave, Gardner, and Musgrave, *The Plant Hunters*, 9–11.

33. Shteir, *Cultivating Women, Cultivating Science.*

34. Blunt, *Travel, Gender and Imperialism*, 46–56.

35. Maddrell, *Complex Locations*, 84.

36. Laura E. Ciolkowski, "Travelers' Tales: Empire, Victorian Travel and the Spectacle of English Womanhood in Mary Kingsley's 'Travels in Africa,'" *Victorian Literature and Culture* 26 (1998): 349.

37. David N. Livingstone, "Tropical Hermeneutics: Fragments for a Historical Narrative: An Afterword," *Singapore Journal of Tropical Geography* 21 (2000): 92–98.

38. Arnold, *The Tropics and the Travelling Gaze*; Daniel P. Miller and Peter H. Reill, eds., *Visions of Empire* (Cambridge: Cambridge University Press, 2010).

39. Felix Driver and Brenda S. A. Yeoh, "Constructing the Tropics: An Introduction," *Singapore Journal of Tropical Geography* 21 (2000): 1–5.

40. Arnold, *The Tropics and the Travelling Gaze*; Stepan, *Picturing Tropical Nature*.

41. Deepak Kumar, *Science and the Raj: A Study of British India* (Oxford: Oxford University Press, 1995), vii.

42. Endersby, *Imperial Nature*; Fa-Ti Fan, *British Naturalists in Qing China* (Chicago: University of Chicago Press, 2004).

43. Livingstone and Withers, *Geographies of Nineteenth-Century Science*, 1–19.

44. Tony Ballantyne, *Orientalism and Race: Aryanism and the British Empire* (Basingstoke: Palgrave, 2002), 39.

45. Londa Schiebinger and Claudia Swan, eds., *Colonial Botany: Science, Commerce and Politics in the Early Modern World* (Philadelphia: University of Pennsylvania Press, 2007).

46. Barbara T. Gates, *Kindred Nature: Victorian and Edwardian Women Embrace the Living World* (Chicago: University of Chicago Press, 1998); Dorothy Middleton, *Victorian Lady Travellers* (London: Routledge and Keegan, 1965); Blunt, *Travel, Gender and Imperialism*.

47. David E. Allen, *The Naturalist in Britain: A Social History* (London: Allen Lane, 1976); Nicolaas Rupke, *Richard Owen: Victorian Naturalist* (New Haven, CT: Yale University Press, 1994).

48. Daniel P. Miller, "Joseph Banks, Empire and 'Centres of Calculation' in Late Hanoverian London," in Miller and Reill, *Visions of Empire*, 21–27; Maddrell, *Complex Locations*.

49. Livingstone, *Putting Science in Its Place*, 45.

50. Vera Norwood, *Made from this Earth: American Women and Nature* (Chapel Hill: University of North Carolina Press, 1993); McEwan, *Gender, Geography and Empire*; Cheryl McEwan, "Gender, Science and Physical Geography in Nineteenth-Century Britain," *Area* 30 (1998): 215–23.

51. Antonia Losano, "A Preference for Vegetables: The Travel Writings and Botanical Art of Marianne North," *Women's Studies* 26 (1997): 423–48; Morin, "Peak Practices."

52. Isla Forsyth, "The More-than-Human Geographies of Field Science," *Geography Compass* 7/8 (2013): 529.

53. Personal communication with Sarah Evans of the Royal Geographical Society (with IBG) regarding female membership of the Royal Geographical Society in 1922.

54. Maddrell, *Complex Locations*, 35.

55. Quoted in Charles Nelson, *Shadow among Splendours: Lady Charlotte Wheeler-Cuffe's Adventures among the Flowers of Burma, 1897–1921* (Dublin: National Botanic Gardens, 2013), 10.

56. Letter from Wheeler-Cuffe to Aunt Cha, February 21, 1897, Parkside, NBG.

57. David N. Livingstone, "Tropical Climate and Moral Hygiene: The Anatomy of a Victorian Debate," *British Journal for the History of Science* 32 (1999): 109.

58. James McGuire and James Quinn, eds., *Dictionary of Irish Biography* (Cambridge: Cambridge University Press, 2009).

59. Nelson, *Shadow among Splendours*, 17–18.

60. James S. Duncan, "The Struggle to be Temperate: Climate and 'Moral Masculinity' in Mid-Nineteenth Century Ceylon," *Singapore Journal of Tropical Geography* 21 (2000): 34–47.

61. Quoted in Nelson, *Shadow among Splendours*, 17.

62. Letter from Wheeler-Cuffe to Mother, July 11, 1897, Colombo, Ceylon, NBG.

63. Nelson, *Shadow among Splendours*, 17–18.

64. Stephen L. Keck, "Picturesque Burma: British Travel Writing 1890–1914," *Journal of Southeast Asian Studies* 35 (2004): 387–414.

65. Leonard Bell, “Not Quite Darwin’s Artist: The Travel Art of Augustus Earle,” *Journal of Historical Geography* 43 (2014): 60–70.

66. Stepan, *Picturing Tropical Nature*, 37.

67. Letter from Wheeler-Cuffe to Mother, July 28 1897, Thayetmyo. NBG.

68. Livingstone, “Tropical Climate,” 93–110; Dane Kennedy, “The Perils of the Midday Sun: Climatic Anxieties in the Colonial Tropics,” in *Imperialism and the Natural World*, ed. John M. McKenzie (Manchester: Manchester University Press, 1990), 118–40.

69. Letter from Wheeler-Cuffe to Mother, July 28 1897, Thayetmyo. NBG.

70. Letter from Wheeler-Cuffe to Aunt Cha, July 29 1897, Thayetmyo. NBG.

71. Johnson, “On the Colonial Frontier.”

72. Alison Blunt, “Imperial Geographies of Home: British Domesticity in India, 1886–1925,” *Transactions of the Institute of British Geographers* 24 (1999): 421–40; Alison Blunt and Robyn Dowling, eds., *Home* (London: Routledge, 2006).

73. Letter from Wheeler-Cuffe to Mother, August 4, 1897, Kyaukpadoung, NBG.

74. Under the British Raj, a syce was a person who took care of horses (a groom), and a chaprassi was a government messenger.

75. Letter from Wheeler-Cuffe to Mother, August 4, 1897, Kyaukpadoung, NBG.

76. Letter from Wheeler-Cuffe to Mother, August 4, 1897, Kyaukpadoung, NBG.

77. Driver and Yeoh, “Constructing the Tropics.”

78. Letter from Wheeler-Cuffe to Mother, August 4, 1897, Kyaukpadoung, NBG.

79. Sonali Jana and G. S. Shekhawat, “Critical Review on Medicinally Potent Plant Species: *Gloriosa superba*,” *Fitoterapia* 82 (2011): 293–301.

80. Letter from Wheeler-Cuffe to Mother, August 4, 1897, Kyaukpadoung, NBG.

81. Letter from Wheeler-Cuffe to Mother, August 4, 1897, Kyaukpadoung, NBG.

82. Letter from Wheeler-Cuffe to Mother, August 4, 1897, Kyaukpadoung, NBG.

83. Duncan, “The Struggle to be Temperate.”

84. James Duncan, “Dis-orientation: On the Shock of the Familiar in a Far-away Place,” in *Writes of Passage: Reading Travel Writing*, ed. James Duncan and Derek Gregory (London: Routledge, 1998), 154.

85. Letter from Wheeler-Cuffe to Mother, September 3, 1897, Thayetmyo, NBG.

86. Letter from Wheeler-Cuffe to Mother, September 3, 1897, Thayetmyo, NBG.

87. Keck, “Picturesque Burma,” 411.

88. Letter from Wheeler-Cuffe to Mother, September 3, 1897, Thayetmyo, NBG.

89. Letter from Wheeler-Cuffe to Mother, September 3, 1897, Thayetmyo, NBG.

90. Letter from Wheeler-Cuffe to Mother, September 25, 1897, Thayetmyo, NBG.

91. Letter from Wheeler-Cuffe to Mother, September 25, 1897, Thayetmyo, NBG.

92. Letter from Otway Wheeler-Cuffe to Charlotte’s Mother, October 24, 1897, Rangoon, NBG.

93. Letter from Otway Wheeler-Cuffe to Charlotte’s Mother, October 24, 1897, Rangoon, NBG.

94. Letter from Wheeler-Cuffe to Mother, November 5, 1897, Thayetmyo, NBG.

95. Letter from Wheeler-Cuffe to Mother, November 27, 1897, Thayetmyo, NBG.

96. Letter from Wheeler-Cuffe to Mother, November 27, 1897, Thayetmyo, NBG.

97. Letter from Wheeler-Cuffe to Father, December 31, 1897, Thayetmyo, NBG.

98. Letter from Wheeler-Cuffe to Father, December 31, 1897, Thayetmyo, NBG.

99. Letter from Wheeler-Cuffe to Father, December 31, 1897, Thayetmyo, NBG.

100. Letter from Wheeler-Cuffe to Mother, January 8, 1898, Thayetmyo, NBG.

101. Nuala C. Johnson, "Global Knowledge in a Local World: Charlotte Wheeler Cuffe's Encounters with Burma, 1901–1902," in *Spaces of Global Knowledge: Exhibition, Encounter and Exchange in an Age of Empire*, ed. Diarmid A. Finnegan and Jonathan J. Wright (London: Ashgate, 2015), 19–38.

102. Livingstone, *Dealing with Darwin*, 1–26, 197–207.

103. Stoler, *Along the Archival Grain*.

CHAPTER EIGHT

Climate, Environment, and the Colonial Experience

VINITA DAMODARAN

British intellectual engagement with the environments of empire involved understanding the colonies as spaces of climatic concern, unfettered opportunity, moral danger, dangerous disease, natural plenitude, and environmental experimentation. South Asia, South Africa, India, Australia, and New Zealand each became giant open-air colonial laboratories, natural archives even, for scientists attempting to understand, investigate, and govern a world of new peoples, species, environments, and diseases. On the one hand, encountering and documenting the facts of environmental change, indigenous knowledge systems, and practices of natural resource use laid the groundwork for much that is modern in environmental thinking, especially in the tropics. Those parts of the Indian subcontinent and sub-Saharan Africa that came under colonial rule provided much of the context for different debates on the relationships between humans and nature, India in particular being something of a template for other parts of the empire in terms of discussions over climate, forestry, environmental history, and colonial governance. On the other hand, colonial encounters and interventions had permanent and uneven consequences on indigenous peoples and their environments.

These things matter, I suggest, because in the modern world, global warming, climate change, extreme climatic events, deforestation, desertification, famine, marine pollution, and soil erosion are considered among the most alarming and important problems threatening humankind. Individually and severally, they represent global challenges normally considered contemporary or recent. Each of them, however, was the subject of attention in the nineteenth century and in different colonial settings. In that period, too, diagnoses and solutions were often offered, some successful, some disastrous, some a prompt to fierce indigenous resistance. The experience of the British and of those colonized by them and the articulation of different colonial environmental narratives in that period is particularly relevant in shaping current debates as well as of intrinsic importance. The human cultures and natural environments of the colonies as written about in the nineteenth century thus assume considerable significance, as do the discourses of the different intellectual communities and scientists who engaged with different colonial peoples, landscapes, and species. Geography matters here as sites and spaces for the making of an innovative science that emerged in the context of colonial scientific and personal networks that crisscrossed empire.[1] The facts of environmental change and "the colonial experience" differed, sometimes fundamentally, often from island to island, certainly from one colonial continental territory to another, as did plans for their amelioration. At the same time, there were important shared circumstances, common influences, and networks of exchange of ideas and knowledge that resulted from collective governance from an imperial metropole, these despite seemingly local specificities in climate, colonial rule, and environmental change.[2]

This chapter offers some thematic arguments—provocations perhaps, rather more than fully documented narratives—within this global picture of climate, empire, and environmental change. Drawing upon evidence from different colonial settings within the British Empire in the nineteenth century, what follows looks at desiccation, narratives of deforestation, meteorology, and soil erosion as environmental and colonial concerns. Recent writers such as David Livingstone have documented the eighteenth- and nineteenth-century colonial debates drawing close and consequential connections among climate, empire, and race—what he calls "climate's moral economy" and "moral climatology."[3] In variously addressing ethnoclimatology, anthropometric cartography, and the links among race, culture, and climate, Livingstone has illuminated the close associations in the nineteenth century between the nexus of climate, race, and discourses of moral judgment on the one hand and their geographical expression on the other. Numerous commentators, in different intellectual contexts and colonial circumstances, "explained" human social difference with reference to the determining influences of climate and pathogenic environments: in short, a degenerate nature produced degenerate natives and so risked the degeneration of European settlers.

Livingstone's construal of the connections among colonial environments, empire, and race is an essentially moral argument. Here, however, I want to illustrate how climate was altered by human agency and how that fact was productive of particular scientific discourses—spaces for debates about forestry, colonial meteorology as a form of imperial science, and debates on environmentalism. This does not suppose a separateness of discourse or a precise chronology in their expression: neither was the case. As Livingstone and others have shown, the co-constituted questions of climate and race in nineteenth-century colonial context were simultaneously moral and medical, social and scientific.[4] Many of the ideas expressed in the 1800s drew upon earlier notions and writers.

In acknowledging these arguments but interpreting the same issues from this different perspective, what follows has two main strands, both (for reasons of space) rather more illustrative than substantive. The first explores how the material realities of climate shaped colonial scientific enterprises in the nineteenth century and how, in different settings, fields such as forestry and meteorology were themselves shaped by, and in, colonial environments and at different scales from the global to the local (often, in a colonial setting, at the level of the island). The second examines the legacy of this work with reference to later scholarship on environmental history. For while it is the case that what is now recognized as global environmental history emerged after 1945, perhaps particularly in the 1960s, it arguably did so in reference to then pressing environmental concerns and at the neglect of important expressions of environmentalism in the nineteenth century. In and from the "environmental decade" of the 1860s especially, studies of climate, environmental degradation, and colonial well-being were shaped by and in different experiences. Practitioners and commentators there and in the metropole shared concerns in common over the growing evidence for the causal and unequal relationship between human agency and an altered nature.[5]

Environmentalism and Environmental History in the Nineteenth Century: Colonial Perspectives

It is noteworthy that several early European writers to comment on environmental change in imperial context were natural philosophers and themselves actors in particular colonial settings. As early as the mid-seventeenth century, several contemporaries were well aware of high rates of soil erosion and deforestation in the colonial tropics, notably on islands. This is true, for example, of Richard Norwood and William Sayle in Bermuda, Thomas Tryon in Barbados, and Edmond Halley and Isaac Pyke on St. Helena.[6] Several of them wrote of the urgent need for, in modern terms, forms of conservationist intervention, especially to protect forests and species threatened by environmental depredation. On St. Helena and in Bermuda, this early conservationist ethic led, by 1715, to the gazetting of the first colonial forest reserves and forest protection laws. On the French colony of Mauritius (the Isle de France), Pierre Poivre and Philibert Commerson likewise framed pioneering forest conservation legislation in the 1760s designed specifically to prevent rainfall decline. In India in the second half of the eighteenth century, the Scottish physician and botanist William Roxburgh wrote what reads now as an almost alarmist narrative on deforestation as a cause of climate change.[7] Together with Alexander Beatson on St. Helena, Roxburgh would speculate upon the incidence of global drought that we know today is prompted by El Niño events.[8]

It was, however, in the particular circumstances of environmental change at the colonial periphery that what we would now term "environmentalism" first made itself felt. In these settings, colonial proponents were often in a position to draw upon firsthand experience and upon historical evidence for environmental change. Whether they were de facto environmental historians because they did so is a moot point. Nevertheless, there was, by the 1840s at least, and in different parts of Britain's colonial world, demonstrable recognition among contemporaries of the connectedness of climate and the human condition. In India, the writings in the 1860s of Scots-born botanist and Indian forester Hugh Cleghorn, and, later in the century, of Berthold Ribbentrop, noted important connections among forests, indigenous practices of management, and the demands of a rapacious empire.[9] In similar fashion for Australia, German botanist Ferdinand von Mueller and, earlier, the Polish-born naturalist and explorer Paweł Strzelecki wrote texts that display an awareness of evidence for global, or, at least, continental, environmental change over time and of the implications for understanding not only the present biotic assemblages then encountered but also for past and, potentially, future plant geographies.[10] With respect to southern Africa, John Croumbie Brown—who wrote also on forestry and human culture in northern Europe and on forest laws in medieval England—was strongly inclined toward the deleterious effects of human agency as he examined the connections among hydrology, forestry, and soil degradation in his *Hydrology of South Africa*.[11] These men and others, such as the American George Perkins Marsh, whose *Man and Nature: Or Physical Geography as Modified by Human Action* (1864) evinced in its subtitle in particular an early ecological consciousness, shared an awareness of the potential human impact on climate and, in particular, the fear that human activity, especially deforestation, might lead to global desiccation.[12]

Evidence from the emerging earth sciences similarly interpreted longer-term climate change with reference to present-day features in the landscape. As Louis Agassiz's theories of ice sheets and progressive Ice Ages grew in acceptance, African explorer David Livingstone recognized lake shoreline features in the Kalahari as signifying high levels of rainfall and events of considerable magnitude in the past and subsequent desiccation.[13] The discovery toward the end of the nineteenth century of moraines on Mt. Kenya and Mt. Kilimanjaro, far below the tongues of present glaciers, was unequivocal evidence that the climate of Africa had been cooler in the past. In West Africa, John Falconer drew upon evidence from earlier commentators in his *Geography and Geology of Northern Nigeria* (1911) to describe extensive fields of linear dunes covered in vegetation and supporting a large human population. These dune fields stretched well into northern Nigeria but far to the south of the then limits of the Sahara.[14]

At much the same time, French explorers under the direction of Jean Tilho brought back evidence of a great lake having existed in the Bodélé Depression between Tibesti and Lake Chad. From this, they reasoned that the Sahara had been more extensive in the past than was now the case and, at other times, less arid.[15] The study of climatic events in the present reinforced these beliefs and emerging theories. By the end of the nineteenth century, it was evident that short-term vicissitudes in the African climate were not only clues to former geographies but that they could have considerable contemporary economic importance. In East Africa, the level of Lake Victoria rose suddenly in 1878 after unusually heavy rains. A few months later, and following prolonged and heavy rainfall over the Blue Nile's catchment in Ethiopia, disastrous flooding of the Nile followed in Egypt. In contrast, southern Africa experienced severe droughts in 1862 and again between 1881 and 1885. In Senegal over the course of the nineteenth century, rainfall declined but showed patterns of biennial variation. As Alexander Knox's 1911 book to that title indicated, studying the climate of the continent of Africa was important for what it revealed about the vital and varying connections between climate and human agency over time.[16] This and other works all spoke to what, essentially, were millennial theories of global desiccation.

I return later on to this desiccationist discourse, to its enduring legacies, and to other expressions of what, now, we can see was a growing environmental historical consciousness. The aforementioned examples may be multiplied many times over, of course, as, in one colonial setting or another, contemporaries recognized the effects of human agency upon climate and the environment. There are grounds for seeing in these cumulative inquiries the beginnings of discrete scientific concerns—what, later in the nineteenth century, would become more clearly defined as disciplines, distinguished by a more particular focus, agreed methods, institutions, and practices. This argument can be made more strongly—namely, that the colonies provided key locales in and from which we can trace the emergence of sciences whose central concerns were a consequence of those locales even as their practitioners turned to matters of pressing global concern. It must not, however, be made too strongly. These locales were not unique in this respect but were, rather, places in which notions of environmental change were either particularly acute or the subject of close attention to colonial or other officials, and sometimes both. Neither is this an account of disciplinary histories when disciplines, sensu stricto, did not exist: it is about discourses whose contents demonstrate an awareness of colonial environmental deterioration as a result of human agency, discourses that in time, and in different settings, would help shape modern subjects both politically and intellectually.

In meteorology—arguably *the* science of climate as established in the colonies—such developments were apparent as early as 1791. As Richard Grove long ago pointed out, that was the year in which "weather and agrarian observations made by scientific observers and others in the tropics were sufficiently elaborate and sufficiently coordinated . . . for some of the first speculations to be firmly made about global rather than regional climatic events."[17] In nineteenth-century India, as Katherine Anderson has shown, the development of meteorology as a form of rational and quantitative inquiry, and, in modern parlance, as a discipline, both reflected the concerns of the imperial state and was itself a politicized form of rational disciplining—in its training of indigenous agents, in establishing standard practices of observation, and so on.[18] In New Zealand, expressions of "environmental anxiety" following the establishment of links among deforestation, soil erosion, rainfall variability, and famine were apparent as early as the 1840s and 1850s.[19] In Britain's colonies in the nineteenth century, the birth of the "modern" fact was attended by concerns over population numbers, disease types, patterns of mortality, and the environmental consequences that lay behind the parallel growth of statistical accountancy in this period.[20]

The term "colonial forestry" or "imperial forestry" should be used with care. It has different connotations: of resources and the language of extractive utilitarianism; a discursive arena in which links were made among it, rainfall, climate, and human well-being; and as a matter of botanical diversity in which colonial tree species were brought to the metropole or to other colonies, or in which European species were introduced to the colonies. Plants and trees were important indicators of climate and acclimatization as moral and environmental matters. As Martin Mahony and Georgina Endfield note, a more strictly scientific forestry and more agreed-upon notions of forest conservation emerged across the colonies of the European empires only toward the end of the nineteenth century and in the first decades of the twentieth century. Discourses about colonial forestry as an imperial revenue scheme were also closely associated with practices of new species introduction.[21] But earlier, from the middle years of the nineteenth century, contemporaries debated the relationships among forest clearance, seasonal or annual variability in rainfall, and, from this, the links among deforestation, climate, and the environment in the longer term.[22]

One illustration must necessarily suffice to make this point. James Fox Wilson's work of 1865, on "the progressive desiccation of inner Southern Africa," was based on an address to the Royal Geographical Society in that year.[23] Wilson was not alone. As Grove has shown, his concerns were part of collective interests in environmentalism and the networks of colonial science in the 1860s among fellows of the Royal Geographical Society and among delegates to section E, Geography, of the British Association for the Advancement of Science. Wilson stressed that desiccation was not natural, "but entirely the consequence of human action." South Africa was his particular concern. The extent and causes of deforestation there required urgent remedial action:

> It being a matter of notoriety . . . that the removal piecemeal of forests, and the burning off of jungle from the summits of hills has occasioned the uplands to become dry and the lowlands to lose their springs . . . it becomes of extreme importance to our South African fellow-subjects, that the destruction of the arboreal protectors of water should be regarded as a thing to be deplored, deprecated, and prevented; and that public opinion on the matter should be educated. . . . But we must not stop there. The evil is one of such magnitude and likely to bear so abundant a harvest of misery in the future, that the authority of law, wherever practicable, should be invoked in order to institute preventive measures.[24]

To Wilson, these facts were everywhere apparent in Britain's colonies: "In our own British colonies of Barbadoes, Jamaica, Penang, and the Mauritius, the felling of forests has also been attended by a diminution of rain."[25] The fact that his views were not shared, including by Africanist David Livingstone, who argued that some of Wilson's plans for reforestation had anyway been adopted and given the slow communication between colony and metropole, meant that Wilson's views did not become the basis for widespread political action. In later decades, similar views did attract political attention through a shared colonial scientific network.

Enduring Legacies: A Partial Geohistoriography of Colonial Environmental History

In the United States in the early twentieth century, several geographers addressed the postglacial desiccation of Central Asia and China based on the twin tenets that wet conditions characterized the glacial phases of the Pleistocene (the first epoch of the Quaternary period), and that aridity had increased since the warming of Pleistocene ice sheets in the Holocene (the current geological epoch, commencing c. 11500 BCE). Travelers in Central Asia pointed to dry watercourses and abandoned settlements as evidence of desiccation and suggested, too, that deteriorating environmental conditions had spurred successive nomadic invasions. A notable figure in this respect was geographer and staunch environmental determinist Ellsworth Huntington, whose travels and intelligence activities in Central Asia powerfully shaped his views.[26] His first major work, *The Pulse of Asia* (1907), set an agenda for both desiccationism and environmental determinism. Both Huntington in the United States and, in Russia, the anarchist Pyotr Kropotkin (the latter in a landmark article published in *The Geographical Journal* in 1904) were influenced by contemporary environmental anxieties in the tropics and a growing interest in climatic interpretations of history.[27] Both also looked to the evidence of the great Indian famines of the late nineteenth century in particular to bolster their arguments over deforestation, environmental change, and negative human consequences.

In Britain, a small group of geographers and others, influenced by contact in the nineteenth century with colonial scientists and geographers, similarly turned to consider the global relations among environmental change, political power, and societal change. Sir Halford Mackinder's *Britain and the British Seas* (1902) was a highly selective historical interpretation of nature, geography, and, to use a modern idiom, superpower political economy.[28] Both this work and Hereford George's *The Historical Geography of the British Empire* (1904) have claims to be considered a world environmental history if judged in relation to the politics of their age. John Linton Myres's insightful work *The Dawn of History* (1911) offered an archaeological and classicist's view of the geographical factor in the rise of ancient civilizations and states: it can be considered de facto a global environmental history of the ancient world.[29] After the hiatus of World War I, further studies linking climate and history appeared, notably Charles Brooks's *The Evolution of Climate* (in 1922) and his *Climate through the Ages* (1926).[30] As a colonial scientist—though he wrote as early as 1918 on the health implications of London fog—Brooks drew conclusions about changes in world climate from colonial territories (where he had been stationed) as far apart as the Falkland Islands and Uganda.[31] Each of these works reflected an awareness of the connections between environmental decline in ancient and contemporary empires and, in Mackinder particularly, strongly held fears over imperial disturbances and the emergence of anti-colonial nationalisms.

Among French scholars, Henri Hubert gave voice to collective desiccationist fears in an influential article in 1924.[32] But it was in the context of German and British colonial semiarid South and Southwest Africa that the gospel of desiccation found its most pronounced and didactic expression after 1918. In 1920, E. H. L. Schwartz's book *The Kalahari; or Thirstland Redemption* gave voice to his views on the progressive desiccation of Africa, their cause and remedy.[33] Where others' earlier work did not reach the ears of legislators, Schwartz's polemical *The Kalahari* did, its message informing the rather alarmist tones of some government documents such as the 1922 report of the South African Drought Commission. That the formal expressions of political concern and institutionalization of these concerns via government legislation were twentieth century in date should not obscure the fact that, for South Africa at least and for specific spaces there, underlying expressions of environmental awareness about the effects of deforestation on desiccation date from the 1860s.

In Africa, concerns expressed in the 1920s turned to embrace colonial territories that had not featured in the earlier environmental literature of the years before 1914 but which, after the war, became the subject of considerable colonial interest and infrastructural investment. One reason for this was that in the 1920s, the Colonial Office was the main employer of British biologists, geologists, topographical surveyors, and geographers. This was especially so in Anglo-Egyptian Sudan, about which colonial context some of the first literature after 1919 on desert spreading or desertification began to be written. A pioneer in this area was the historian of Africa E. W. Bovill, who echoed Schwartz in South Africa in a 1921 paper on "the encroachment of the Sahara on the Sudan."[34] Bovill's work was taken up by G. T. Renner in one of the first articles to depict Africa as a potentially famine-ridden continent.[35]

Across the Atlantic, the emergence of a period of environmental alarmism in North America consequent on the prolonged "Dustbowl" droughts in the southern United States in the early 1930s supplemented the existing colonial panic over desert spreading.[36] This increased local anxieties and affected policy in some British and French colonies. The New Deal conservationism of the United States was emulated in the east and central African colonies in particular. Soil erosion had already become prominent as an issue in India from the 1890s, and huge investments were made to control it, in the Etawah region of the United Provinces of northern India, for example. It is noteworthy that these efforts, like their counterpart in west and in southern Africa, long predated the American "Dustbowl" alarmism. In 1936–1937, Edward Stebbing, prominent Indian forester and longtime professor of forestry at the University of Edinburgh, visited northern Nigeria and Niger as part of an Anglo-French expedition examining the effects of deforestation on desertification. By the time of his visit, Stebbing had already published a three-volume work titled *The Forests of India*, much of which detailed the history of environmental concerns and early conservationism among the first surgeon-foresters of the East India Company medical service.[37] His detailed knowledge of Indian issues and his encounter with French and British West African colonies—significantly, it took place during the dry season—prompted his rather apocalyptic warning over what he knew to be the dangers of desert spreading.

The title of his essay, "The Encroaching Sahara: The Threat to the West African Colonies," suggests a familiarity with Bovill's similarly titled work of 1921.[38] Colonial scientists who collectively had greater experience with the causes, rates, and seasonality of local desertification and erosion downplayed Stebbing's warnings. Yet his concerns, as for Wilson in the 1860s, were the subject of detailed debate among geographers and others in the metropole and, unlike Wilson's, were taken up by governing circles in Paris and in London.[39] Stebbing's work led directly to the founding of the Anglo-French Boundary Forest Commission in 1934. In 1937, Stebbing published his concerns at greater length in *The Forests of West Africa and the Sahara*.[40] A year later, his somewhat alarmist warnings were considered to be largely unjustified, and his analysis was dismissed by a commission member, Brynmor Jones.[41] Undaunted, Stebbing went on to publish a further deforestation-desiccation-desertification narrative in 1953.[42]

If, then, debates about desiccation, deforestation, and desertification in Britain's African colonies were the subject of concern, albeit not without dissenting voices from a few, from the 1860s and more heatedly from the 1920s and 1930s, we may also note parallels—and differences—in other colonial settings. In Australia, and in Australian geography, connections between climate and settlement were the concern of Griffith Taylor. His work on aridity, settlement, and race is sometimes overread by later commentators as a rather virulent environmental determinism, but, in the context of his time—Taylor was involved with the visit of the British Association to Australia in 1914, for example—it was born of exploring what was possible in expanding the frontiers of human (colonial) settlement in Australia.[43] Others expressed the problems more forcefully. In 1937, Francis Ratcliffe, who had studied the causes of soil erosion in South Australia and Queensland, published *Flying Fox and Drifting Sand*, a harsh indictment of the impact of extensive outback agriculture. This went through numerous editions and, in its emphasis on the degradation of the Australian environment through capitalized agriculture and pastoralism, exercised a disproportionate influence on global environmental concerns far outside the antipodean context.[44] Ratcliffe drew upon the writings of Keith Hancock, later a prominent historian of the British Empire and Commonwealth, who in 1931 published an attack on profligate deforestation and land clearing by settlers.[45] Interestingly, Hancock, a Quaker, was the biographer of Jan Smuts and was undoubtedly influenced by the "holistic" views of the latter, not least as an early environmental campaigner and environmental historian, a point that Hancock noted in his 1972 work, one of the earliest and most influential texts in Australian environmental history.[46]

In the United States, Graham Jacks and R. O. Whyte's *The Rape of the Earth: A World Survey of Soil Erosion* marshaled evidence for the kind of rapid environmental change predicted by John Croumbie Brown and George Perkins Marsh seventy years before, and did so with reference to detail that such earlier figures had not been in a position to compile.[47] This book helped set the scene for the postwar British (and French) colonial obsession with soil erosion and gullying in their "second colonial occupations," as well as for the global desertification "mania" of the 1970s in the wake of major droughts in West Africa.

By and large, these later post-1945 histories are not my concern, although scholars are now tracing how cadres of colonial scientists retooled themselves in the postwar age of decolonization and, via the United Nations, were pivotal in building notions of "the" global environment as an object to be studied, mapped, conserved, and ameliorated.[48] Here, I hope to have shown how, by the mid-1920s and 1930s, there was an innovative convergence of analytical and descriptive writings by geographers, ecologists, anthropologists, and archaeologists even, over the environmental anxieties promoted by human-environment interactions. Much of their data was nineteenth century in origin and colonial in its geographical setting. Their geographical focus and their scientific networks were with Britain's colonies and, increasingly, with the global effects and scale of environmental change resulting from human action. Yet what also distinguished this work was something approaching a sort of individual and institutional "amnesia" concerning the environmental debates of the second half of the nineteenth century. Only from the mid-1950s and in the 1960s was the work of the many overlooked environmentalists of a century before or even earlier exhumed and brought to contemporaries' attention. Two leading figures in doing so were the geographers David Lowenthal, and, in a different way, Clarence Glacken. Both are important for their attention to the debates on climate, race, and environment: Lowenthal for his engagement with the work of George Perkins Marsh as a proto-ecological figure in the nineteenth century and Glacken for a work that outlines the long history of human-nature attitudes in the western intellectual imagination but that ends its account in 1800. Assessment of their work is, therefore, a fitting coda to this necessarily truncated geo-historiographical survey.

In *Versatile Vermonter*, his biography of George Perkins Marsh, Lowenthal highlighted the way in which Marsh utilized accounts of environmental degradation in the Mediterranean world, the decline of Roman and other empires, and the degradation of colonial island colonies to offer warnings about the contemporary unsustainable use of resources in the United States, particularly on its then colonial western frontier.[49] Marsh, argued Lowenthal, knew the lessons to be learned from the past: Marsh's *Man and Nature* evidenced a global history of environmental change and an acquaintance with colonial literatures (especially French and British) on the imperial ecological impact (as did Lowenthal's own work). Where Lowenthal looked to Marsh and others in the 1860s to reflect on the intellectual origins of conservationism, Clarence Glacken took an altogether *longue durée* view in his magisterial *Traces on the Rhodian Shore* in 1967, subtitled *Nature and Culture in Western*

Thought, from Ancient Times to the End of the Eighteenth Century.[50] Glacken's first book, *The Great Loochoo,* based on his doctoral dissertation, was a historical geography of Okinawa.[51] For Glacken, this work emphasized the historical significance of insular environmental constraints, so that the intellectual jump from consideration of island to world—profoundly a matter of scale, geographically and in terms of interpretation—seemed logical, as it had to Marsh a century earlier. Glacken's *Traces* might never have been completed had it not been for Carl Sauer's intellectual protection of him from the critique of skeptical colleagues at Berkeley. Sauer, the doyen of American cultural geography and human ecology and with his own interests in the global history of plant domestication and cultural landscape evolution, saw Glacken's work as vital in making the history of environment ideas global. Glacken's book encompassed thinking about nature from the Akkadian to the European maritime empires, exploring in particular the emergence of ideas associated with humankind as *the* dominant environmental agent.

Although at first prevented by its title from receiving an appropriate popular reception, Glacken's *Traces* almost immediately spun off several works inspired by his overview of global environmentalism. The work of Russell Meiggs titled *Trees and Timber in the Ancient World* is notable in this respect.[52] Donald Hughes further pursued the ecological history, proto-environmentalism, and the alleged ecologically driven demise of classical Greece and Rome in his *The Environmental History of the World.*[53] Following his reading of Glacken, English intellectual historian Keith Thomas wrote *Man and the Natural World* (1983), a work that usefully filled in some of the gaps left by Glacken in his treatment of English-based environmental thinking.[54] But, like Glacken, Thomas ended his account at 1800, leaving a lacuna in environmental history for the whole of the nineteenth century, which, arguably, has not been properly treated by any environmental historian either for Britain, the United States, or, indeed, the world as a whole. Simon Schama's *Landscape and Memory* acknowledges Glacken but is confined almost entirely to an inward-looking Europeanism.[55] This ignores the fact that Glacken's key contribution was to recognize not only that European attitudes to nature could not be understood without reference to European expansion and colonial settlement but that they were essentially a part of a global intellectual evolution. Outside the Anglo-American context, environmental historians inspired mainly by extra-European history and/or influenced by the French *Annales* school have also credited Glacken as their inspiration. In 1987, Spanish historian Luis Urteaga published *La Tierra Esquilmada,* a history of Spanish environmental attitudes and policies during the eighteenth century.[56] Glacken is also prominent in the work of Brazilian José Pádua on the proto-environmentalism of early nineteenth-century Latin America.[57] Other disciples of Glacken, such as John McNeill and Joachim Radkau, have studied environmental history in the twentieth century.[58] Their grounds for doing so is that the turning point in modern environmental history occurred in the twentieth rather than in the nineteenth century, a view this chapter has sought to dispel.

Conclusion

This chapter has sought to explore environmental history in colonial context, the sites and spaces of an emerging science of the environment, and to examine how colonial environmental thinking and scientific networks were apparent in debates in the nineteenth century and how those debates were, and were not, engaged with by later scholars. Later work, I have suggested, was an emergent global history of environmental change based at least in part upon an understanding of the colonial experience of ecological impact and environmental change. Partial and perhaps provocative though it is, this chapter goes someway to recovering forgotten or overlooked literature whose legacy is apparent but whose originating context is less so.

Other conclusions may be drawn, some in a firmer hand, others more lightly. This perspective on climate and colonialism complements and does not contradict Livingstone's work. It reinforces the value of his attention to the sites and spaces of science's making and reception, whether of particular settings such as botanic gardens or Empire Forestry conferences, or of colonial spaces at the scale of islands (St. Helena) or continents (Australia). It highlights the importance of thinking and working comparatively, over time, over space, and by theme, about environmental history and the colonial experience. Particular topics—insofar as they may be easily discerned from one another—may be examined by theme, in given colonial settings, or as an emergent discourse-cum-discipline. But we must remember that the colonial experience did not occur at one scale and that it has left an unequal archival legacy.

This is true in several senses. In each of these colonial settings, the natural environment there is, in general terms, its own archive, painstakingly revealed through dendrochronology, relict physical features, stratifications in desert sediment, pollen analysis, and so on. Present understanding of these past geographies, even if only partial, is vital in providing analogues for our future. Even as this is so, the result of site-specific investigation is on occasion used, at times unthinkingly, to stand for something larger than itself, "up-scaled" without recognition of the importance of geographical difference. Further, cultural meanings in the landscape do not so easily leave enduring traces, not least when indigenous inhabitants have been diminished in number or detached from a long-run "rootedness" in their environment by the exigencies of empire. It is true, too, that the paper trails and documentary stratification left in "the colonial archive" do not necessarily reflect fully what happened so much as they speak to different practices of record keeping and, usually, to the authority of metropolitan views over the voices and views of those in the periphery.[59] This is not to subscribe unthinkingly to the view, discussed by Thomas Richards in his *The Imperial Archive*, that much metropolitan "rendering" of empire in its archives was fictive, more fantasy than a record of actual governance.[60] It is to argue that rich possibilities remain in archival colonial records for understanding how and why, in the nineteenth century, climate and environment became the subjects of concern in the ways they did and at the scales they did, and how historians, either of paper or of nature, resident either in metropole or colony, might learn from one another as they seek to understand them and their geographical expression.

Notes

1. David Lambert and Alan Lester, *Colonial Lives across the British Empire: Imperial Careering in the Long Nineteenth Century* (Cambridge: Cambridge University Press, 2006).

2. In a voluminous literature on these topics, the work of environmental historian Richard Grove is particularly important. See Richard Grove, "Origins of Western Environmentalism," *Scientific American* 267 (1992): 42–47; Richard Grove, *Green Imperialism: Colonial Expansion, Tropical Island Edens and the Origins of Environmentalism* (Cambridge: Cambridge University Press, 1995); Richard Grove, *Ecology, Climate and Empire: Colonialism and Global Environmental History 1400–1940* (Cambridge: White Horse Press, 1997); Richard Grove, Vinita Damodaran, and Satpal Sangwan, eds., *Nature and the Orient: The Environmental History of South and Southeast Asia* (Cambridge: Cambridge University Press, 1998). Grove's foundational work in establishing modern environmental history (he established the journal *Environment and History*) is evident in the Festschrift volume marking his work: Deepak Kumar, Vinita Damodaran, and Rohan d'Souza, eds., *The British Empire and the Natural World: Environmental Encounters in South Asia* (Oxford: Oxford University Press, 2011). For other work of similar scope, see Alfred Crosby, *Ecological Imperialism: The Biological Expansion of Europe, 900–1900* (Cambridge: Cambridge University Press, 1986); David Arnold and Ramachandra Guha, eds., *Nature, Culture, Imperialism: Essays on the Environmental History of South Asia* (Delhi: Oxford University Press, 1995); Mahesh Rangarajan, "Environmental

Histories of South Asia," *Environment and History* 2 (1996): 129–43; Richard Grove and Vinita Damodaran, "Environment," in *The Ashgate Research Companion to Modern Imperial Histories*, ed. Philippa Levine and John Marriott (Aldershot: Ashgate, 2012), 567–80; Gregory Barton, "Empire Forestry and the Origins of Environmentalism," *Journal of Historical Geography* 27 (2001): 529–52; and Gregory Barton, *Empire Forestry and the Origins of Environmentalism* (Cambridge: Cambridge University Press, 2002).

3. David N. Livingstone, "The Moral Discourse of Climate: Historical Considerations on Race, Place, and Virtue," *Journal of Historical Geography* 17 (1991): 413–34; David N. Livingstone, "Climate's Moral Economy: Science, Race and Place in Post-Darwinian British and American Geography," in *Geography and Empire*, ed. Anne Godlewska and Neil Smith (Oxford: Blackwell, 1994), 132–54; David N. Livingstone, "Tropical Climate and Moral Hygiene: The Anatomy of a Victorian Debate," *British Journal for the History of Science* 32 (1999): 93–110; David N. Livingstone, "Race, Space and Moral Climatology: Notes toward a Genealogy," *Journal of Historical Geography* 28 (2002): 159–80; David N. Livingstone, "Tropical Hermeneutics and the Climatic Imagination," *Geographische Zeitschrift* 90 (2002): 65–88; David N. Livingstone, "Cultural Politics and the Racial Cartographics of Human Origins," *Transactions of the Institute of British Geographers* 35 (2010): 204–21.

4. For an excellent overview of these issues, see Martin Mahony and Georgina Endfield, "Climate and Colonialism," *Wiley Interdisciplinary Reviews: Climate Change* 9 (2018): https://doi.org/10.1002/wcc.510.

5. These issues are identified in summary fashion in Richard Grove and Vinita Damodaran, "Imperialism, Intellectual Networks, and Environmental Change: Origins and Evolution of Global Environmental History, 1676–2000: Part I," *Economic and Political Weekly* 41 (2006): 4345–54. This chapter builds upon this earlier work, and our 2012 essay, Grove and Damodaran, "Environment," and I acknowledge with thanks the steer of the coeditors toward David Livingstone's work in doing so.

6. See, for example, *Memorials of the Discovery and Early Settlement of the Bermudas or Somers Islands, 1515–1685: Compiled from the Colonial Records and Other Original Sources, by Major-General J. H. Lefroy* (London: Longmans, Green, and Company, 1877–79); Thomas Tryon, *Friendly Advice to the Gentlemen-Planters of the East and West Indies: In three parts* by Philotheos Physiologus [alias Thomas Tryon] (London: Printed by Andrew Sowle, 1684). On Halley and others on St. Helena, see Grove, *Green Imperialism*, 114.

7. Grove, *Green Imperialism*, 399–408.

8. Richard Grove, "Revolutionary Weather: The Climatic and Economic Crisis of 1788–1795 and the Discovery of El Niño," in *A Change in the Weather: Climate and Culture in Australia*, ed. Tim Sherratt, Tom Griffiths, and Libby Robin (Canberra: National Museum of Australia Press, 2005), 128–40; Richard Grove, "The East India Company, the Raj and the El Niño: The Critical Role Played by Colonial Scientists in Establishing the Mechanisms of Global Climate Teleconnections 1770–1930," in Grove, Damodaram, and Sangwan, *Nature and the Orient*, 301–23.

9. Hugh F. Cleghorn, *The Forests and Gardens of South India* (Edinburgh: Oliver and Boyd, 1861). On Cleghorn's pioneering work in Indian forestry, see Henry J. Noltie, *Indian Forester, Scottish Laird: The Botanical Lives of Hugh Cleghorn of Stravithie* (Edinburgh: Royal Botanic Garden, 2016). See also Berthold Ribbentrop, *Forestry in*

British India, 2 vols. (Calcutta: Office of the Superintendent of Government Printing, 1900).

10. [Baron] Ferdinand von Mueller produced a range of works on the indigenous flora of Australia. Important in this context are his *Report on the Forest Resources of Western Australia* (London: Reeve, 1879); *Plants of North-Western Australia* (Perth: By authority of R. Pether, Government Printer, 1881); and *Systematic Census of Australian Plants, with Chronologic, Literary and Geographic Annotations* (Melbourne: Printed for the Victoria Government by McCarron, Bird and Company, 1889). On Mueller's contribution to botanical studies in nineteenth-century Australia, see Edward Kynaston, *A Man on Edge: A Life of Baron Sir Ferdinand von Mueller* (London: Allen Lane, 1981). On Strzelecki, see his *Physical Description of New South Wales and Van Diemen's Land* (London: Longman, Brown, Green, and Longmans, 1845). For this book and for the explorations that underlay it, Strzelecki was awarded the Gold Medal of the Royal Geographical Society in 1846.

11. John C. Brown, *The Hydrology of South Africa* (Kirkcaldy: Crawford, 1875).

12. For the European intellectual context of these studies of colonial environmental change, see Paul Warde, *The Invention of Sustainability: Nature and Destiny, circa 1500–1870* (Cambridge: Cambridge University Press, 2018), 265–349.

13. Louis Agassiz, *Études sur les Glaciers: Ouvrage accompagné d'un Atlas de 32 planches* (Neuchâtel: Jent et Gassmann, 1840); David Livingstone and Charles Livingstone, *Narrative of an Expedition to the Zambesi and Its Tributaries: And of the Discovery of the Lakes Shirwa and Nyassa, 1858–1864* (New York: Harper and Bros., 1866).

14. John D. Falconer, *The Geography and Geology of Northern Nigeria* (London: Macmillan, 1911).

15. Jean Tilho, *Documents Scientifiques de la Mission Tilho, 1906–1909*, 2 vols. (Paris: Imprimerie Nationale, 1910–11).

16. Alexander Knox, *The Climate of the Continent of Africa* (Cambridge: Cambridge University Press, 1911).

17. Grove, "The East India Company, the Raj and the El Nino," in Grove, Damodaram, and Sangwan, *Nature and the Orient*, p. 302.

18. Katherine Anderson, *Predicting the Weather: Victorians and the Science of Meteorology* (Chicago: University of Chicago Press, 2005).

19. James Beattie, "Environmental Anxiety in New Zealand, 1840–1941: Climate Change, Soil Erosion, Sand Drift, Flooding and Forest Conservation," *Environment and History* 9 (2003): 379–92; James Beattie, Emily O'Gorman, and Matthew Henry, eds., *Climate, Science, and Colonization: Histories from Australia and New Zealand* (Basingstoke: Palgrave Macmillan, 2014).

20. Mary Poovey, *A History of the Modern Fact: Problems of Knowledge in the Sciences of Wealth and Society* (Chicago: University of Chicago Press, 1998); Theodore Porter, *The Rise of Statistical Thinking, 1820–1900* (Princeton, NJ: Princeton University Press, 1986).

21. Mahony and Endfield, "Climate and Colonialism." On the later institutionalism of "empire forestry," for example, see J. M. Powell, "'Dominion over Palm and Pine': The British Empire Forestry Conferences, 1920–1947," *Journal of Historical Geography* 33 (2007): 852–77.

22. See Barton, *Empire Forestry*.

23. See James Fox Wilson, "On the Progressing Desiccation of the Basin of the Orange River in Southern Africa," *Proceedings of the Royal Geographical Society* 9 (1864–1865): 106–9, and his "Water Supply in the Basin of the River Orange, or 'Gariep, South Africa," *Journal of the Royal Geographical Society* 35 (1865): 106–29. On this work, see Richard Grove, "A Historical Review of Early Institutional and Conservationist Responses to Fears of Artificially Induced Global Climate Change: The Deforestation-Desiccation Discourse 1500–1860," *Chemosphere* 29 (1995): 1001–13 (Grove here miscites Wilson's work as that of James Spotswood Wilson [1013]), and Nancy Jacobs, *Environment, Power and Injustice: A South African History* (Cambridge: Cambridge University Press, 2003), 61–64.

24. Wilson, "On the Progressing Desiccation," 108; On Wilson's views, see Richard Grove, "Imperialism and the Discourse of Desiccation: The Institutionalisation of Global Environmental Concerns and the Role of the Royal Geographical Society," in *Geography and Imperialism*, ed. Morag Bell, Robin A. Butlin, and Michael Heffernan (Manchester: Manchester University Press, 1995), 36–52.

25. Wilson, "On the Progressing Desiccation," 107–8.

26. For his principal publications in this respect, see Ellsworth Huntington, *The Pulse of Asia: A Journey in Central Asia, Illustrating the Geographic Basis of History* (Boston: Houghton Mifflin and Company, 1907); *Civilization and Climate* (New Haven, CT: Yale University Press, 1915); *Climatic Changes: Their Nature and Causes* (New Haven, CT: Yale University Press, 1922); *Principles of Human Geography* (New York: Wiley, 1922).

27. Pyotr Kropotkin, "The Desiccation of Eur-Asia," *Geographical Journal* 23 (1904): 722–41.

28. Halford J. Mackinder, *Britain and the British Seas* (London: Heineman, 1902). On Mackinder's geopolitical work and his environmental perspectives, see Gerry Kearns, *Geopolitics and Empire: The Legacy of Halford Mackinder* (Oxford: Oxford University Press, 2009), 69–71, 81–84, 151–53.

29. John L. Myres, *The Dawn of History* (London: Williams and Norgate, 1911).

30. Charles E. P. Brooks, *The Evolution of Climate* (London: Benn, 1922); *Climate through the Ages: A Study of the Climatic Factors and their Variations* (London: Benn, 1926).

31. Charles E. P. Brooks, *Climate and Weather of the Falkland Islands and South Georgia* (London: Meteorological Committee, 1920); *Variations in the Levels of Central African Lakes, Victoria and Albert* (London: Meteorological Committee, 1923).

32. Henri Hubert, "Le dessèchement progressif en Afrique occidentale française," *Bulletin du comité d'études historiques et scientifiques de l'Afrique occidentale française* 17 (1920): 11–24.

33. E. H. L. Schwartz, *The Kalahari; or Thirstland Redemption* (Oxford: Blackwell, 1920).

34. E. W. Bovill, "The Encroachment of the Sahara on the Sudan," *Journal of the Royal African Society* 20 (1921): 23–45.

35. G. T. Renner, "A Famine Zone in Africa: The Sudan," *Geographical Review* 16 (1926): 583–96.

36. On this, see Donald Worster, *Dust Bowl: The Southern Plains in the 1930s* (New York: Oxford University Press, 1979).

37. Edward Stebbing, *The Forests of India*, 3 vols. (London: Bodley Head, 1922–1926).

38. Edward Stebbing, "The Encroaching Sahara: The Threat to the West African Colonies," *Geographical Journal* 85 (1935): 506–19.

39. See Percy Fox et al., "The Encroaching Sahara: The Threat to the West African Colonies," *Geographical Journal* 85 (1935): 519–24, and Francis Rodd, J. D. Falconer, and E. P. Stebbing, "The Sahara," *Geographical Journal* 91 (1938): 354–59.

40. Edward Stebbing, *The Forests of West Africa and the Sahara: A Study of Modern Conditions* (London: W. and R. Chambers, 1937).

41. Brynmor Jones, "Desiccation and the West African Colonies," *Geographical Journal* 91 (1938): 401–23.

42. Edward Stebbing, *The Creeping Desert in the Sudan and Elsewhere in Africa: 15°–13° Latitude* (Khartoum: McCorquodale and Company, 1953).

43. J. M. Powell, "Thomas Griffith Taylor, 1880–1963," *Geographers Biobibliographical Studies* 3 (1979): 141–54. On Griffith Taylor's own view of his writings in this respect, see T. Griffith Taylor, *Journeyman Taylor: The Education of a Scientist* (London: Robert Hale, 1958), especially chaps. 16 and 17.

44. Francis Ratcliffe, *Flying Fox and Drifting Sand: The Adventures of a Biologist in Australia* (London: Chatto and Windus, 1937).

45. W. Keith Hancock, *Australia* (New York: C. Scribner's and Sons, 1931).

46. W. Keith Hancock, *Discovering Monaro: A Study of Man's Impact on His Environment* (Cambridge: Cambridge University Press, 1972); J. M. Powell, "Signposts to Tracts: Hancock and Environmental History," unpublished paper given to a symposium at the Australian National University, Canberra, April 1–3, 1998.

47. Graham V. Jacks and R. O. Whyte, *The Rape of the Earth: A World Survey of Soil Erosion* (London: Faber and Faber, 1939).

48. Perrin Selcer, *The Postwar Origins of the Global Environment: How the United Nations Built Spaceship Earth* (New York: Columbia University Press, 2018).

49. David Lowenthal, *George Perkins Marsh: Versatile Vermonter* (New York: Columbia University Press, 1958).

50. Clarence Glacken, *Traces on the Rhodian Shore: Nature and Culture in Western Thought, from Ancient Times to the End of the Eighteenth Century* (Berkeley: University of California Press, 1967).

51. Clarence Glacken, *The Great Loochoo: A Study of Okinawan Village Life* (Berkeley: University of California Press, 1955).

52. Russell Meiggs, *Trees and Timber in the Ancient Mediterranean World* (Oxford: Clarendon Press, 1982).

53. Donald Hughes, *The Environmental History of the World: Humankind's Changing Role in the Community of Life* (London: Routledge, 2001).

54. Keith Thomas, *Man and the Natural World: Changing Attitudes in England 1500–1800* (London: Allen Lane, 1983).

55. Simon Schama, *Landscape and Memory* (London: Harper Collins, 1995).

56. Luis Urteaga, *La tierra esquilmada: Las ideas sobre la conservación de la naturaleza en la cultura española de siglo XVIII* (Barcelona: Ortega, 1987).

57. José Pádua, *Sopro de destruição: Pensamento político e crítica ambiental no Brasil escravista (1786–1888)* (Rio de Janeiro: Casimiro, 2002).

58. John M. McNeill, *Something New under the Sun: An Environmental History of the World in the Twentieth Century* (London: Allen Lane, 2000), and Joachim Radkau, *The Age of Ecology* (Cambridge: Polity Press, 2014).

59. On this point and particularly for the colonial context, see Ann Laura Stoler, *Along the Archival Grain: Thinking through Colonial Ontologies* (Princeton, NJ: Princeton University Press, 2008).

60. Thomas Richards, *The Imperial Archive: Knowledge and the Fantasy of Empire* (London: Verso, 1993).

CHAPTER NINE

Lost in Place

Two Expeditions Gone Awry in Africa

DANE KENNEDY

One of the most significant outcomes of the spatial or geographical "turn" in the history of science has been a new appreciation of the vital role that travel and exploration played in the production of scientific knowledge. As David N. Livingstone noted in *Putting Science in Its Place*, "scientific knowledge in Europe depended on global circulation, and domestic maps of knowledge were continually reoriented in the light of the faraway."[1] From the moment of first contact with the Americas and other overseas lands, the "faraway" would exert an ever-increasing influence on the direction, pace, and scale of European scientific investigation, giving a double meaning to the notion of scientific discovery.

Francis Bacon recognized in the early seventeenth century that ship logs and travel journals offered valuable information for insights into the natural world. By the end of the seventeenth century, the Royal Society and other learned bodies were drafting instructions and sponsoring publications that sought to discipline travelers and systematize their observations for scientific purposes. The relationship between travel and science grew even more intimate and institutionalized in the second half of the eighteenth century as Britain, France, and other European states sent specially equipped vessels into the Pacific to map its blank spaces and gather meteorological data, botanical specimens, and other scientific information about this vast, unfamiliar region. By the early nineteenth century, newly founded organizations such as the Royal Geographical Society and the British Association for the Advancement of Science were promoting exploration as a scientific endeavor that required a new type of traveler, a technically proficient, quasi-professional figure whose arrival on the scene necessitated a neologism: the explorer.[2]

The information gathered from travel proved particularly useful for advances in subjects such as astronomy (especially when applied to navigation), botany, cartography, ethnography, geography, hydrology, and meteorology—what Steven Harris has termed the "big sciences" because of their scale and variability. Yet the physical challenges of travel to distant lands made it difficult to "do" science in a way that measured up to the standards set by metropolitan authorities. "How science travels," Harris reminds us, "has as much to do with the problem of travel in the making of science as it does with the problem of making science travel."[3] Not only did travel present physical and logistical challenges that scientists ensconced in their libraries and laboratories did not face, but it also produced its own particular set of epistemological problems. Insofar as the verification of scientific discoveries rested on replication, this was much more difficult to do when those discoveries were made in distant places under difficult circumstances. Moreover, the crucial if intangible issue of trust, which has come to be seen as integral to the rise of modern science, proved especially problematic for those who conducted journeys of discovery. As Steven Shapin has argued in his influential study of the gentleman scientists of seventeenth-century England, one way to establish trust was to be part of a shared community, a tight-knit social network.[4] Yet most travelers and explorers were outsiders in the eyes of the scientific community, and the long history of tall travelers' tales that Jonathan Swift satirized so memorably in *Gulliver's Travels* weighed heavily against their truth claims.

To win the trust of scientific gatekeepers, travelers and explorers had to find new ways of supplying proof of their observations and accomplishments. This is what Livingstone refers to as new "modes of scientific knowing."[5] One important mode consisted of explorers' embodied experiences: their own physical suffering during long and arduous journeys became a privileged site of knowledge and source of authority.[6] They also made more systematic use of scientific instruments to map and to measure their travels, thereby turning the particularities of place into information that could contribute to a universal knowledge.[7] Finally, they recognized that their scientific credibility and reputation were enhanced by their involvement in imperial agendas. The famed eighteenth-century voyages that Britain's James Cook, France's Louis-Antoine de Bougainville, Spain's Alejandro Malaspina, and various others made on behalf of their respective countries brought science into service of the state, using the exploration of the natural world as an instrument of empire. According to Bruno Latour, the very structure of modern science came to be organized around imperial "centers of calculation," each exerting its own field of gravity. Even the grammar of scientific exploration became infused with imperial overtones. As one historian of science has noted: "Travelling naturalists, concerned with the global distribution of flora and fauna, relied upon the infrastructure of colonial expansion and borrowed the language of imperialism to describe plant and animal geography."[8] Empire, in effect, gave exploration greater scientific legitimacy.

Historical geographers and others have devoted increased attention in recent years to exploration as a manifestation of imperial science in motion. They have produced important studies of the scientific work carried out by exploration's luminaries, such as Britain's Captain James Cook, Roderick Murchison, and David Livingstone, as well as by a number of lesser-known agents.[9] While these studies acknowledge the challenges and setbacks that such figures faced, they are mainly concerned with their accomplishments. Yet plenty of explorers faced more than setbacks: they met with disaster, their efforts achieving little or nothing. What happens when the "problem of travel" proves insurmountable and leads to the collapse of the endeavor? What does it mean for a scientific expedition to fail? And what does that failure tell us about place as a constituent element in the pursuit of scientific knowledge?

I raise these questions in the context of two expeditions that the British government sent into the interior of Africa in the immediate aftermath of the Napoleonic Wars. Both were large, expensive, grandly ambitious endeavors. Both were conducted, at least in part, in pursuit of what contemporaries construed as important scientific questions. Yet both expeditions were unmitigated failures. Both were soon forgotten, erased from the annals of exploration. Even so, much can be learned from them about the entanglements of science in state-sponsored efforts to expand the geographical reach of Britain's commercial and imperial influence, as well as the political, logistical, and microbial constraints that local conditions placed on those efforts. The story of these expeditions puts the spatiality of early nineteenth-century science not simply in its place, as David Livingstone has so memorably put it, but in a place where the problem of travel undermined scientific knowledge itself.

The Congo Expedition

In July 1815, just a month after the Napoleonic Wars had been brought to a bloody but triumphant end at the battle of Waterloo, British authorities devised a grand strategy to simultaneously solve what European geographers regarded as two of Africa's greatest geographical mysteries. Where did the Congo River begin and the Niger River end? If the British could answer those questions, they could open the interior of Africa to British commerce and impose British imperial influence over the region, unimpeded by French interference. The Admiralty launched a naval expedition up the Congo River that sought to bypass its barrier of cataracts and continue to its source. The War and Colonial Office organized an army expedition designed to march inland from the Windward coast, establish contact with African states in the interior, and follow the Niger River to its outlet. Some geographers suspected that the Niger River and the Congo River were one and the same, which raised the prospect of the two expeditions meeting each other in the interior. These expeditions were conceived by their planners, then, as partner endeavors, conducted concurrently and potentially conjoined. Taken together, they comprised what may have been the most ambitious exploratory initiative the British ever launched in Africa.

John Barrow, the Admiralty's second secretary and a prominent proponent of African exploration, organized the Congo expedition. Although Barrow would acquire a reputation over the next few decades as Britain's leading impresario of state-sponsored expeditions, this was the first large undertaking to be launched under his direction.[10] He made two early decisions that indicate the importance he attached to the expedition's scientific objectives. The first was to appoint Captain James Hingston Tuckey as the expedition's commander. Tuckey had spent several years conducting coastal surveys in New South Wales. On his return voyage to Britain in 1805, he was captured by French forces and languished in a French prison for the next eight and a half years. He spent his time in incarceration writing an encyclopedic, four-volume study of *Maritime Geography and Statistics*, which was published in early 1815, little more than a year after his release from prison. Barrow saw Tuckey as a man who possessed the intellectual curiosity, technical skills, and prior experience required to lead a scientific expedition.[11] It is worth noting that Tuckey refers only once in *Maritime Geography* to the Congo River: it "is seldom visited by ships," he observed, "and is consequently little known."[12]

Barrow's other key decision was to solicit the advice of Sir Joseph Banks, the eminent naturalist and president of the Royal Society. Now crippled by gout and nearing the end of his life, Banks—who had originally made his name some forty-five years earlier as the naturalist on Captain James Cook's first Pacific voyage—was thrilled by Barrow's invitation. "Your letter has made my old blood circulate with renewed vigour," he enthused. Banks had a long-standing interest in Africa, and as a founding member of the Association for Promoting the Discovery of the Interior Parts of Africa (commonly known as the African Association), he had sponsored a series of expeditions into Africa, including Mungo Park's famous journey to the upper reaches of the Niger River in 1795–1797. Expressing "the pleasure I derive from finding our Ministers mindful of the Credit we have obtained from Discovery," Banks no doubt saw an opportunity to make the Congo expedition a worthy African successor to the Pacific voyages of Cook, Vancouver, and others.[13] What this required, first and foremost, was a suitably trained naturalist who could oversee a team of scientific specialists. After some delay, Banks found his man—Dr. Chetian Smith, a Norwegian who had studied botany with Professor J. W. Hornemann at the University of Copenhagen and who had recently returned from an expedition to Madeira and the Canary Islands with the geologist Baron von Buch. To assist Smith, Banks also obtained the services of David Lockhart, a gardener at Kew Gardens. The other members of the scientific team were John Cranch, who collected marine specimens for the British Museum; William Tudor, a comparative anatomist from Liverpool; and Edward Galwey, a gentleman of independent means who had taught himself a modicum of chemistry, botany, and geology. Galway had appealed to Captain Tuckey, a personal friend, to accompany the expedition at his own expense, "plead[ing] the example of Sir Joseph Banks," who had done the same on the Cook expedition.[14]

Banks exerted his influence on the expedition's preparations in another important, but more surprising, way: he urged the Admiralty to commission the construction of a steam vessel to take the expeditionary party up the Congo River. As he saw it, only a steamboat could effectively overcome the Congo's strong current. In addition, it could serve as "a fort impregnable to Native Armies" if, as he expected, local peoples proved hostile.[15] To appreciate what a radical recommendation this was, it is important to understand that steam propulsion technology was still in its infancy. Only eight years earlier, Robert Fulton had established the first commercial steamboat service on the Hudson River in New York, followed several years later by a similar service on the Clyde River in Scotland. The first steamboat plied the Thames River only a few months prior to Banks's proposal—perhaps inspiring the idea—and none had yet attempted an oceanic voyage.[16] Barrow was skeptical of the practicality of steam power, as was Tuckey, but Banks wielded so much influence in elite circles that his recommendation could not be ignored. Consequently, the Admiralty contracted with the Birmingham engineering firm of Boulton and Watt, the premier manufacturer of steam engines, to design and build a steamship for use by the Congo expedition at an estimated cost of £1,700.

According to historian of technology Daniel Headrick, "no single piece of equipment is so closely associated with the idea of imperialism as is the armed shallow-draft steamer."[17] The Congo expedition marks the first attempt to bring the steamer into use for such purposes. James Watt promised to build a boat that would steam up the Congo at seven knots with a maximum four-foot draft. When the vessel was completed, however, Admiralty engineers found that its speed was far slower and draft much deeper than anticipated. Even with its masts and stores removed, the newly christened "Congo" only reached five and a half knots and still failed to achieve its draft specifications. "I am now convinced," Barrow declared, "that a Steam Engine would be useless and pernicious to the Navigation of the Congo." He added, "I was always of this opinion . . . but I was overruled." He was overruled by a person he does not name, though he obviously had Banks in mind.[18] The "Congo" was soon refitted for the expedition as a conventional sailing vessel, thus bringing to a close this ambitious if premature scheme to apply one of the most innovative technologies of the day to the exploration of Africa.

With the expedition ready to depart, Barrow issued official letters of instruction to Captain Tuckey and the leading members of his scientific team. These detailed documents specified the roles and responsibilities of their recipients and in so doing provide the fullest statement of the scientific goals of the expedition. It is impossible to describe here the exceptional specificity of these instructions, but Innes Keighren, Charles Withers, and Bill Bell have rightly noted that "if carried out to the letter, such instructions . . . seemingly left little time to do anything other than measuring, observing, and noting down, rather than writing at length."[19] Yet writing at length was an expectation as well: Barrow urged Tuckey and his scientists to keep detailed journals in triplicate. The expedition's main task was to trace the course of the Congo River and determine whether it was, in fact, the Niger—which Tuckey, unlike Barrow, thought highly unlikely.[20] Tuckey was required as he moved upriver to fix his longitude and latitude at regular intervals, and to measure the river's depth, the strength of its current, and the rise and fall of water levels. He was urged also to seek out possible river routes to the east coast of Africa. Meanwhile, information about the region's climate, mountains, animals, vegetables, minerals, peoples, languages, and other topics was to be collected and recorded. The scientific members of the expedition received instructions that meticulously detailed not only what they should do but how they should do it. One of William Tudor's tasks was to capture that "singular animal," a hippopotamus, preserve its heart, lungs, and entrails, and determine how "it contrives to keep itself . . . at pleasure under water, and to walk upright at the bottom of the River."[21] Barrow and the Admiralty had high expectations indeed.

The Congo expedition set sail in February 1816. It consisted of two ships—the "Congo" and the "Dorothy," which carried supplies and trade goods—a crew of forty-seven, the five scientific "supernumeraries," and two African translators of self-professed Congolese origin. (One really was Congolese; the other was not.) The expedition was well stocked with scientific instruments and supplies. The instruments included two pocket chronometers, a sextant, a quadrant, an artificial horizon, a reflecting telescope, four thermometers, two mountain barometers, two hygrometers, a microscope, three pocket compasses, two azimuth compasses, one Walker's compass, one pocket telescope, and one measuring chain; the supplies included a blowpipe, syringes, spirits, preserving power, packing cases, bottles, labels, pens, and paper. In June 1816, the expedition reached the mouth of the Congo, where Tuckey and his crew would soon confront a daunting set of challenges.

The Niger Expedition

Lord Bathurst, secretary of War and Colonies, and his undersecretary, Henry Goulburn, were Barrow's counterparts in planning the Niger expedition. The War and Colonies Department had little experience with scientific exploration of the sort the Admiralty had been conducting for half a century in the Pacific and elsewhere. It had, however, sponsored Mungo Park's second expedition to West Africa in 1805. This attempt to trace the course of the Niger had ended disastrously with the deaths of Park and his forty-five-man party, most of whom died of disease, though Park himself evidently drowned in the river. The new expedition was intended to accomplish the goals that Park had failed to achieve. Its aim was to march overland from the West African coast to the headwaters of the Niger and then to trace the river's course to its conclusion. Convinced that this endeavor required a force of sufficient size and power to impress or intimidate African rulers, Bathurst approved an expeditionary party that consisted of sixty-nine Royal African Corps troops (forty of them white, twenty-nine black), thirty-two African civilian personnel, and a substantial number of camp followers. A caravan of nearly two hundred pack and draft animals was obtained to transport rations and supplies, along with several field cannon, an arsenal of other weapons and munitions, and an abundance of gifts.[22] The entire enterprise bore more than a passing resemblance to an invading army.

Major John Peddie, an army officer with African experience, was appointed to lead the expedition, with Captain Thomas Campbell of the Royal Staff Corps his second-in-command. The British West African colony of Sierra Leone served as the expedition's staging ground, and its governor, Lieutenant Colonel Charles MacCarthy, supplied political intelligence, advice, and manpower. While scientific personnel and concerns loomed less large in this expedition than in the Congo venture, they were not negligible to its preparations. Adolphus Kummer, a Parisian-trained naturalist from Saxony who had originally come to Senegal with French forces, was recruited as a specimen collector and draughtsman. Staff Surgeon William Cowdrey's talents were said to include some knowledge of astronomy and mineralogy, a familiarity with Arabic, and experience as an explorer in South Africa. Peddie reported that the expedition had taken on a "Gentleman" named Mr. Ross, who was said to "understand Navigation perfectly," though he does not appear again in the expedition records.[23] Captain Campbell made the astronomical observations needed to chart the route and location. The expedition was equipped with a whole raft of scientific instruments, including theodolite, barometers, thermometers, and chronometers.

Peddie's official letter of instruction was not quite as detailed as the one Barrow drafted for Tuckey, but its expectations were clear. The opening sentence declared the purpose of the expedition to be "the Improvement of the Geographical Knowledge of the Interior of Southern Africa and eventually to the Extension of British Commerce in that Quarter." "The Main Object of the Expedition," it continued, "is the discovery of the Mouths of the Niger and the Course of that River." The letter held open the possibility that the Niger and the Congo were one and the same river, offering hope of the Tuckey expedition "aiding your progress in the latter Stages of your Enterprize." Peddie was ordered to survey his course and "fix as far as you may be able the Latitude and Longitude" of notable locations along the way, as well as observe the "general Character, Religion, Customs, Manners and the principal Products of the Country." The instructions also placed emphasis on the need to "pay every deference to the feelings and prejudices of the Inhabitants," recognizing that the success of the expedition depended on their cooperation.[24] Further evidence of the importance the expedition's planners placed on encounters with African communities came in the form of a supplementary memorandum drafted by Thomas Harrison, secretary of the African Institution, which oversaw missionary operations in Sierra Leone. He detailed the types of information the expedition should seek to collect about the region, its peoples, and their practices. (A copy of the report was forwarded to Barrow for use by the Congo expedition under the naive expectation that it would encounter indigenous societies much the same as those familiar to Harrison.[25])

The Niger expedition suffered several major setbacks even before it got underway. Crowdey, the expedition's most scientifically able officer, succumbed to disease shortly after reaching the base camp on the coast of Senegal. Peddie appointed Duncan Dochard, a young doctor with the Royal African Corps in Freetown, to replace him. A week later, Peddie himself was dead. Captain Campbell assumed command of the expedition. At least seven other men died before the party set out from the coast in February 1817. Fatalities declined as the caravan marched inland, but disease continued to cause problems. It would prove even more devastating to the Congo expedition, giving ample support to Africa's popular reputation as the "white man's grave." This was a place with a disease environment that proved profoundly hostile to the physical well-being of the individuals who carried out these early nineteenth-century scientific endeavors.

Deadly microbes, however, were hardly the only challenges that place posed for the expeditionary parties as they ventured into the African interior. The Africans who inhabited these places also had much to say about the success or failure of the two expeditions.

Africans as Intermediaries and Impediments

Both expeditions sought the cooperation of the Africans whose territories they passed through. Both recruited African intermediaries to serve as guides, translators, collectors, and informants. Both depended on African communities for labor, shelter, food, and supplies. This meant that Africans were not simply strange creatures whose customs, practices, and beliefs made them the objects of scientific study. They were also active contributors to the scientific work of the expeditions, though often in ways that were unacknowledged. The contributions they made were grounded in their intimate understanding of the places the expeditions passed through, thus exemplifying one of the ways exploration put "science in its place."

This reliance on African agents raises serious questions, however, about the provenance of the scientific knowledge obtained by explorers. Their claims of scientific authority derived above all from their direct observations in the field, an "ocular demonstration" of evidence that so-called armchair geographers and other metropolitan experts could not provide.[26] Yet those observations invariably occurred through various forms of mediation by African assistants. These assistants guided expeditionary parties through unfamiliar territory, served as translators for communications with local peoples, collected plants and caught insects, imparted information about surrounding regions and their inhabitants, and much more. Their contributions to expeditions have attracted increasing attention from scholars in recent years.[27] Both the Congo and the Niger expeditions relied on indigenous intermediaries. Somme Simmons, one of the two ex-slaves who accompanied the Congo expedition from England, proved especially useful to its operations. His father happened to be the ruler of Boma, the main trading port along the lower Congo River, and in gratitude for the return of his son, he overruled those local merchants who were hostile to the British, offering his cooperation instead. Simmons himself led the team of local guides that accompanied the expedition as it ventured farther upriver. The Niger expedition's principal guide was Adrian Parterriaux, a Soninke trader at the mouth of the Senegal River who possessed a range of professional and personal contacts in the interior. In addition, many of the African soldiers who volunteered to participate in the expedition were ex-slaves from the territories it intended to pass through, and their knowledge of the local languages and customs were important assets.

Both expeditions, then, relied mainly on African intermediaries who had been victims of the slave trade. They were not alone in doing so: the British resorted repeatedly to what David Lambert has aptly termed "enslaved knowledge" in their efforts to learn more about the West African interior.[28] The disruptive effects of the slave trade were also felt by the two expeditions in ways that proved far less advantageous to their purposes. Because Britain was engaged at this time in an active campaign to suppress the slave trade, Africans whose livelihoods depended on that trade tended to view British explorers with suspicion and antagonism. The Boma merchants who opposed any assistance to the Congo expedition did so because they feared it intended to stop their traffic in human beings. One of them "broke out into a violent passion, abusing and calling [the king of England] 'the devil.'" James Tuckey had to give local elites "assurances of not coming to prevent the slave trade, or to make war."[29] Even so, slave merchants threw repeated obstacles in the expedition's way.

Tensions over the slave trade proved even more troublesome for the Niger expedition. The route it took to reach the upper Niger River required it to pass through Futa Jallon, an Islamic state that engaged in the slave trade and opposed Britain's efforts to suppress it. Despite efforts by Peddie and his successor Campbell to placate the Almaami, Futa Jallon's ruler, this powerful figure was understandably suspicious of this heavily armed party's intentions. Further complicating matters, Futa Jallon was at war with the neighboring non-Muslim state of Kaarta over issues aggravated by the slave trade. Although the Almaami repeatedly reassured Captain Campbell of his friendly intentions, he engaged in a systematic campaign to delay and disrupt the Niger expedition's passage through his territory.

His efforts were aided by the catastrophic collapse of the expedition's logistical capabilities. The horses, bullocks, camels, and donkeys that the expedition had imported from the Cape Verde islands and elsewhere to transport its supplies died off at an alarming rate. Campbell details in his diary the deadly impact of microbes, parasites, poisonous plants, lions, and even killer bees on his steadily shrinking stock. He tried to hire local porters, but in doing so, he merely placed his party at the mercy of the Almaami, who skillfully used the supply and withdrawal of porters to make extortionate demands for payment and to inhibit the forward movement of the expedition. Campbell had to cast off nonessential goods and equipment: even the expedition's prized field guns were abandoned, though buried to keep them from falling into African hands.[30] For months, no progress was made. Finally, Campbell ordered his bedraggled party to retreat to the coast. No sooner had it done so than Campbell, his second-in-command, and the naturalist Kummer died of fever or dysentery.

Rather than acknowledge their failure and draw useful lessons from the experience about the particular challenges posed by the places they sought to penetrate, the British simply reconstituted the expedition under the command of Major William Gray and sanctioned another attempt, this time from a staging ground a hundred miles north of the original point of departure. Once again, the expedition relied on a large train of imported animals to move its supplies, and, as before, the beasts quickly died off. Once again, the expedition needed local porters, and, as before, the ruler of the state whose territory it was passing through—in this case the Almaami of Bondu—used the party's dependency to drain it of resources and control its movement.

The Almaami worked, as Gray noted, "to oppose our further progress."[31] As had earlier been the case, an enterprise premised on continuous forward movement found itself trapped in place.

The Congo expedition ran into logistical problems as well. Knowing that his sailing ship would only take him a short distance up the Congo River, Tuckey had the foresight to insist that the expedition bring two "double-boats" (essentially catamarans, which he may have become familiar with during his time in the South Pacific). Stable, mobile, and propelled by both paddle and sail, these boats carried Tuckey, the team of naturalists, and a dozen or so members of the crew as far as the cataracts, which begin some ninety miles from the river's mouth. At this point, however, the party had to dismantle their boats and march overland.

This landward turn led to new logistical problems for the expedition. It could only advance with the assistance of local porters and access to local supplies. Yet the former were unreliable and the latter scarce. As obstacles mounted, an increasingly frustrated Tuckey gave vent to his feelings in his diary. "The progress of civilization," he concluded darkly, "can only be done by colonization."[32] Simmons and several native guides evidently concluded that the expedition was a lost cause and slipped out of camp in the dead of night. Men began to fall ill. It soon became clear that the party would have to turn back. Tuckey described the return journey as "worse to us than the retreat from Moscow."[33] By the time they reached the "Congo" and "Dorothy" and set sail for England, nearly two dozen members of the expedition, including Tuckey and all of the naturalists except for Kew horticulturalist David Lockhart, had died, most of them the victims of yellow fever.

Assessing Failure

Neither the Congo nor the Niger expedition came close to achieving its main scientific purpose, which was to trace the course of the African river that each was named after. The Niger expedition not only failed to follow the Niger River to its egress; it never even reached the river's headwaters, something Mungo Park had accomplished on his first expedition several decades earlier. Similarly, the Congo expedition not only failed to find the source of the Congo River; it never even got past the cataracts, the same barrier that had blocked Portuguese and other European explorers for centuries. It would take another decade for the brothers Richard and John Lander to prove from firsthand observation that the Niger River flowed into the Gulf of

Guinea. It would be another *six* decades before Henry Morton Stanley, the Anglo-American explorer, traced the Congo River from its origins to the sea.

James Secord has argued in his agenda-setting essay, "Knowledge in Transit," that historians of science need to give closer attention to the "communicative action" that occurs as scientific knowledge is transmitted by means of notebooks, lab registers, museum catalogues, travel books, and other media to its audiences or consumers.[34] By this measure, too, the two expeditions were failures. Few traces of their efforts to collect information entered into the communication circuits through which new discoveries traveled and gained purchase as scientific knowledge.

Although the two iterations of the Niger expedition lasted from 1817 to 1821, those five years generated remarkably little in the way of botanical specimens, meteorological data, astronomical readings, or other information of interest to the scientific community at home. Whatever samples of flora and fauna the naturalist Kummer may have collected do not appear to have survived the journey: they are likely to have been discarded as pack animals died off and conditions grew more desperate. The only traces of his scientific work that remain are a few maps and topographical sketches, along with a brief journal that appears to have attracted no subsequent interest. Although the expedition's successive commanders, Campbell and Gray, kept reasonably detailed journals, they focused almost exclusively on logistical and political matters, offering little that would have been of interest to the scientific community. Campbell's journal never appeared in print, though Gray drew on it in the opening chapters of his own account of the expedition, published by John Murray in 1825. A reviewer of the book pointedly complained that Gray rarely described the places he had visited and failed to give longitude and latitude readings of their locations.[35] The expedition's most useful scientific information came from Staff Surgeon Dochard. In a medical report, he observed that large doses of "the Bark" (quinine) "in no instance failed to effect a cure" among African troops with fevers, though the Europeans sometimes required a dose of calomel as well.[36] Later, he noted having "recourse to the remedies made use of by the natives of Africa, and whenever those were resorted to in time, the disease soon gave way."[37] Regrettably, he did not explain what that disease or those remedies were. Whatever influence Dochard might have had on the treatment of diseases endemic to West Africa was lost when he died soon after the second iteration of the expedition.

The scientific expectations surrounding the Congo expedition were far higher and the disappointment of its sponsors consequently far greater. "No enterprize, since the voyages of Cook, excited a greater share of public interest than that of Captain Tuckey to Explore the river Congo," pronounced John Barrow in his postmortem on the expedition. He added that the captain and his crew "may be said to have fallen victims of a too ardent zeal in the pursuit of science."[38] That zeal was made manifest in the posthumous publication of Tuckey's and Smith's journals as *Narrative of an Expedition to Explore the River Zaire* (1818), which revealed their intellectual curiosity, strength of will, and, especially in the case of Tuckey, powers of observation. Barrow wrote a lengthy introduction to the volume, defending the expedition's scientific objectives and paying homage to its lost leaders. He also included seven appendices on information gathered by the expedition. These included reports on the zoological and botanical specimens and geological samples that had made it back to Britain, along with some hydrographical data and vocabulary lists for two Congolese languages.[39] The apparent purpose of the appendices was to demonstrate that the costly venture had not been in vain. But was anyone persuaded? It is telling that Barrow himself would make no mention of the Congo venture in his autobiography.[40]

Finally, the disastrous outcomes of the Congo and Niger expeditions should be considered in the context of the cultural and political meanings the British attached to failure in the nineteenth century. Stephanie Barczewski has recently noted that the British exhibited a peculiar enthusiasm for "heroic failure," admiring acts of "physical courage and mental fortitude in the face of defeat." The individuals who exemplified these noble characteristics were invariably agents of empire, and explorers such as Mungo Park, John Franklin, David Livingstone, and Robert Falcon Scott held prominent places in this pantheon of heroes. Barczewski suggests that the celebration of heroic failure permitted "Britons to see their empire in a positive and moral light, particularly at moments when the reality was quite different."[41]

There were no such celebrations of James Tuckey or John Peddie or Thomas Campbell or the others members of the Congo and Niger expeditions who suffered and died in those ordeals. Why not? The answer may lie in part in the connection Barczewski draws between exploration and empire. Not only did the Congo and Niger expeditions fail to advance British imperial ambitions in the regions they sought to penetrate; they set back those ambitions in significant and lasting ways. The British never established an imperial presence in the Congo basin. They did so in the Niger basin, but only near the end of the century, long after the futility of the Peddie/Campbell/

Gray expedition(s) had faded from memory. These were failures of political purpose, but they carried scientific consequences. As an imperial science, exploration's scientific accomplishments were inextricably associated with the advancement of empire. The Congo and Niger expeditions left little to celebrate in either regard.

How Place Mattered

The Congo and Niger expeditions dampened British imperial ambitions but did not extinguish them. John Barrow went on to make a name for himself as the driving force for exploration at the Admiralty. He sponsored a series of high-profile expeditions over the following decades, directed mainly at the Arctic, rather than at Africa.[42] Although the War and Colonial Office supported further efforts to explore the West African interior, it turned to the North African state of Tripoli as the point of departure for explorers such as Hugh Clapperton and Dixon Denham. Exploration as conceived and carried out by the British continued to operate in tandem with science. Many of the Victorian era's most eminent scientists—Charles Darwin, Thomas Huxley, and Joseph Hooker are famous examples—began their careers as naturalists on expeditions overseas.

The technological dimensions of science also retained a close association with exploration. The siren call of the steamship, which had so enraptured Joseph Banks, continued to appeal to those who wanted to penetrate the African interior and open it to commerce, Christianity, and colonialism. In the early 1830s, a group of Liverpool merchants commissioned the construction of two steam-powered vessels to go up the Niger River and establish trade with the peoples along its banks. Nearly every member of the forty-man crew fell victim to fevers. In 1841, the British government invested heavily in another attempt to penetrate the Niger, building three specially designed steamships for the purpose. Disease again brought a premature end to the project. It was not until Dr. William Balfour Baikie confirmed in the course of his 1854 expedition up the Niger that regular doses of quinine warded off malaria that the steamboat became a viable instrument of European entry into the Niger basin. The other great African river systems proved even more problematic for penetration by steamer. British authorities placed high hopes in the Livingstone-led expedition they sent up the Zambezi River in 1857, but their purpose-built steamboat repeatedly ran aground in the river's shallows and came to a complete halt when it reached impassible cataracts. Steamboats on the upper Nile found it impossible to penetrate the Sudd, the mass of floating vegetation that clogged the river as it snaked its way through South Sudan. The Congo's cataracts remained an insuperable barrier: not until a railway line was constructed to bypass them toward the end of the century did steamboats begin to ply the river above Stanley Pool.[43] Africa's topography repeatedly exposed the technological hubris that helped inspire exploration as a scientific enterprise.

Yet the British did succeed in exploring much of Africa over the course of the nineteenth century, and in so doing they gathered a great deal of geographical, botanical, geological, and other scientifically useful information about the continent. Whether they drew any direct lessons from the failures of the Congo and Niger expeditions is hard to say, but subsequently the most successful expeditions were those that pursued a very different strategy of engagement with the continent and its peoples. Nearly all of these expeditions were small-scale operations, led by one or two explorers who employed teams of indigenous guides, translators, porters, and other auxiliary personnel. Hugh Clapperton, Heinrich Barth, David Livingstone, Richard Burton, John Hanning Speke, and Verney Lovett Cameron were leading exemplars of this strategy. Their expeditions invariably made a virtue of necessity, relying heavily on local peoples for food, shelter, and various other forms of assistance. In Muslim-majority regions, explorers often adopted local dress and sometimes sought to disguise their identities. Almost everywhere, they assumed conciliatory stances and accommodated themselves as best they could to local customs.

These practices usually eased an expedition's passage through unfamiliar territory, but they also placed limitations on the scientific work it could accomplish. Unable to assemble a team of experts in surveying, botany, zoology, geology, meteorology, and the like, the individual explorer had to acquire a smattering of knowledge about the various technical subjects that mattered to sponsors at home. Hence the proliferation of publications that provided explorers and other travelers with advice and introductions to various branches of science. When Richard Burton explored the lake region of East Africa in 1857–1859, he brought along heavily annotated copies of Julian Jackson's *What to Observe; or, the Traveller's Remembrancer* (1841), John Herschel's *A Manual of Scientific Enquiry; Prepared for the Use of Officers in Her Majesty's Navy, and Travellers in General* (1849), and Francis Galton's *Art of Travel; or, Shifts and Contrivances Available in Wild Countries* (1855).[44] Other African explorers relied on such publications as well. They became, Burton complained, "overworked professional[s]" who were "expected to survey and observe, to record meteorology and trigonometry, to shoot and stuff birds and beasts, to collect geological specimens and theories," and more.[45] Needless to say, their efforts often failed to meet standards set by specialists at home, giving rise to repeated controversies over the accuracy and significance of their findings. These complications were the inevitable outgrowth of the constraints that conditions in Africa placed on the composition of these expeditions. Place mattered in scientific practice. Nowhere was this truth more readily apparent than in the exploration of Africa.

Notes

1. David N. Livingstone, *Putting Science in Its Place: Geographies of Scientific Knowledge* (Chicago: University of Chicago Press, 2003), 143. Livingstone also provides an impressive *tour d'horizon* of the spatial turn that has informed influential thinkers in a range of disciplines in "The Spaces of Knowledge: Contributions towards a Historical Geography of Science," *Environment and Planning D: Society and Space* 13 (1995): 5–34. See also Steven Shapin, "Placing the View from Nowhere: Historical and Sociological Problems in the Location of Science," *Transactions of the Institute of British Geographers* 23 (1998): 5–12.

2. Jason Pearl, "Geography and Authority in the Royal Society's Instructions for Travelers," in *Travel Narratives, the New Science, and Literary Discourse, 1569–1750*, ed. Judy A. Hayden (London: Routledge, 2012), 71–85; Charles W. J. Withers, *Placing the Enlightenment: Thinking Geographically about the Age of Reason* (Chicago: University of Chicago Press, 2007); Charles W. J. Withers, *Geography and Science in Britain, 1831–1939: A Study of the British Association for the Advancement of Science* (Manchester: Manchester University Press, 2010); Felix Driver, *Geography Militant: Cultures of Exploration and Empire* (Oxford: Blackwell, 2001); Dane Kennedy, ed., *Reinterpreting Exploration: The West in the World* (New York: Oxford University Press, 2014). As Craciun has shown, the term "explorer" was commonly a back-projection, ascribed to the author (often by himself [*sic*]) as he, upon safe return, prepared in-the-field writings for publication. See Adrianna Craciun, "What Is an Explorer?" *Eighteenth-Century Studies* 41 (2011): 29–51.

3. Steven J. Harris, "Long-Distance Corporations, Big Sciences, and the Geography of Knowledge," *Configurations* 6 (1998): 271.

4. Steven Shapin, *A Social History of Truth: Civility and Science in Seventeenth-Century England* (Chicago: University of Chicago Press, 1994).

5. Livingstone, *Putting Science in Its Place*, 41, 42.

6. Dorinda Outram, "On Being Perseus: New Knowledge, Dislocation, and Enlightenment Exploration," in *Geography and Enlightenment*, ed. David N. Livingstone and Charles W. J. Withers (Chicago: University of Chicago Press, 1999), 281–94; Livingstone, *Putting Science in Its Place*, esp. chaps. 2 and 4.

7. Marie-Noëlle Bourguet, Christian Licoppe, and H. Otto Sibum, eds., *Instruments, Travel and Science: Itineraries of Precision from the Seventeenth to the Twentieth Century* (London: Routledge, 2002); Fraser MacDonald and Charles W. J. Withers, eds., *Geography, Technology and Instruments of Exploration* (Farnham: Ashgate, 2015).

8. Bruno Latour, *Science in Action: How to Follow Scientists and Engineers through Society* (Milton Keynes: Open University Press, 1987); Diarmid A. Finnegan, "The Spatial Turn: Geographical Approaches to the History of Science," *Journal of the History of Biology* 41 (2008): 381.

9. Brian W. Richardson, *Longitude and Empire: How Captain Cook's Voyages Changed the World* (Vancouver: University of British Columbia Press, 2005); Robert A. Stafford, *Scientist of Empire: Sir Roderick Murchison, Scientific Exploration and Victorian Imperialism* (Cambridge: Cambridge University Press, 1989); Lawrence Dritsas, *Zambezi: David Livingstone and Expeditionary Science in Africa* (London: I. B. Tauris, 2010); Harry Liebersohn, *The Traveler's World: Europe to the Pacific* (Cambridge, MA: Harvard University Press, 2006); David Lambert, *Mastering the Niger: James MacQueen's African Geography and the Struggle over Atlantic Slavery* (Chicago: University of Chicago Press, 2013); D. Graham Burnett, *Masters of All They Surveyed: Exploration, Geography, and a British El Dorado* (Chicago: University of Chicago Press, 2000).

10. Fergus Fleming, *Barrow's Boys* (New York: Atlantic Monthly Press, 1998).

11. John Barrow explains his reasons for appointing Tuckey in letters to Lord Melville and Sir Joseph Banks. See Barrow to Melville, August 9, 1815, in Sir John Barrow letterbooks, LBK/65/2, Caird Library, National Maritime Museum, and Barrow to Banks, August 5, 1815, D.T.C.19.169, Banks Correspondence, Natural History Museum Library and Archives.

12. James Hingston Tuckey, *Maritime Geography and Statistics, or a Description of the Ocean and Its Coasts, Maritime Commerce, Navigation*, 4 vols. (London: Black, Perry, and Company, 1815), 2:541.

13. Banks to Barrow, July 30, 1815, Barrow letterbooks.

14. [John Barrow], "Narrative of an Expedition. . . ," *Monthly Review* 18 (January 1818): 361–62.

15. Banks to Barrow, August 12, 1815, Barrow letterbooks.

16. Ben Marsden and Crosbie Smith, *Engineering Empires: A Cultural History of Technology in Nineteenth-Century Britain* (Houndmills: Palgrave Macmillan, 2005), chap. 3.

17. Daniel R. Headrick, *The Tools of Empire: Technology and European Imperialism in the Nineteenth Century* (New York: Oxford University Press, 1981), 18.

18. Barrow to Home Popham, January 29, 1816, Barrow letterbooks.

19. Innes M. Keighren, Charles W. J. Withers, and Bill Bell, *Travels into Print: Exploration, Writing, and Publishing with John Murray, 1773–1859* (Chicago: University of Chicago Press, 2015), 42–43.

20. Tuckey to Barrow, August 30, 1815, Barrow letterbooks.

21. Instructions to Mr. Tudor, February 7, 1816, Barrow letterbooks. Copies of the instructions to Tuckey, Smith, Cranch, and Lockhart can be found here as well.

22. For lists of the expedition's men and animals, see CO2/5/79, 80, British National Archives (BNA), Kew.

23. John Peddie to Lord Bathurst, March 16, 1816, CO2/5/30–32, BNA.

24. The instructions are reproduced in full in Bruce L. Mouser, "Introduction," in *The Forgotten Peddie/Campbell Expedition into Fuuta Jaloo, West Africa, 1815–17* (Madison: University of Madison African Studies Publications, 2007), 15–17.

25. Thomas Harrison to Barrow, December 24, 1815, with enclosures, Barrow letterbooks.

26. This issue is addressed by Charles W. J. Withers, "Mapping the Niger, 1798–1832: Trust, Testimony and "Ocular Demonstration" in the Late Enlightenment," *Imago Mundi* 56 (2004): 170–93.

27. See Simon Schaffer, Lissa Roberts, Kapil Raj, and James Delbourgo, eds., *The Brokered World: Go-Betweens and Global Intelligence, 1770–1820* (Sagamore Beach: Science History Publications 2009); Felix Driver and Lowri Jones, *Hidden Histories of Exploration: Researching the RGS-IBG Collections* (London: Royal Holloway, University of London, 2009); Shino Konishi, Maria Nugent, and Tiffany Shellam, eds., *Indigenous Intermediaries: New Perspectives on Exploration Archives* (Acton: Australian National University, 2015); Adrian S. Wisnicki, "Charting the Frontier: Indigenous Geography, Arab-Nyamwezi Caravans, and the East African Expedition of 1856–59," *Victorian Studies* 51 (2008): 103–37; Adrian S. Wisnicki, "Victorian Field Notes from the Lualaba River, Congo," *Scottish Geographical Journal* 129 (2013): 210–39.

28. Lambert, *Mastering the Niger*, 4. See also Dane Kennedy, *The Last Blank Spaces: Exploring Africa and Australia* (Cambridge, MA: Harvard University Press), 167–74.

29. James H. Tuckey and Christen Smith, *Narrative of an Expedition to Explore the River Zaire, Usually Called the Congo, in South Africa, in 1816* (London: John Murray, 1818), 266, 110.

30. Journal books of Captain Thomas Campbell, CO2/5/116–489, BNA.

31. Major William Gray and Staff Surgeon Dochard, *Travels in Western Africa in the Years 1818, 19, 20, and 21 from the River Gambia through Woollii, Bondoo, Galam, Kasson, Kaarta, and Foolidoo to the River Niger* (London: John Murray, 1825), 212.

32. Tuckey and Smith, *Narrative*, 187.

33. Tuckey and Smith, *Narrative*, 222.

34. James A. Secord, "Knowledge in Transit," *Isis* 95 (2004): 652–72.

35. See the review of Gray's *Travels* (almost certainly by John Barrow) in *Monthly Review* 108 (1825): 18–20.

36. Dr. Dochard, "Report on the Diseases which Have Prevailed amongst the Men Composing the Expedition for the Interior of Africa" (January 19, 1817), CO 2/5/89–93, BNA.

37. Gray and Dochard, *Travels*, 140.

38. [Barrow], "Narrative," 335, 336.

39. Tuckey's original journal manuscript can be found in ADM 1/2617, BNA, while Smith's Danish-language journal is cataloged as Add MS 32444, British Library. The most substantial body of material to be brought back from the Congo consisted of some six hundred botanical specimens. They were presented to Joseph Banks, who passed them on to Robert Brown, secretary of the Linnaean Society, for identification and classification. See Dr. Richard G. Hingston, "Captain Tuckey of the Congo" (unpublished book manuscript, 2002), appendix A. A copy of Hingston's study is available at the British Library. The journals of John Cranch, the expedition's marine zoologist, were rediscovered—in the Natural History Museum of Paris, curiously enough—and published as Theodore Monod, *John Cranch, Zoologiste de l'Expedition du Congo* (London: Bulletin of the British Museum, 1970).

40. John Barrow, *An Autobiographical Memoir of Sir John Barrow, Bart., Late of the Admiralty* (London: John Murray, 1847).

41. Stephanie Barczewski, *Heroic Failure and the British* (New Haven, CT: Yale University Press, 2016), 20, 226.

42. Fergus Fleming, *Barrow's Boys* (New York: Atlantic Monthly Press, 1998).

43. Kennedy, *Last Blank Spaces*, 135–38.

44. Dane Kennedy, *The Highly Civilized Man: Richard Burton and the Victorian World* (Cambridge, MA: Harvard University Press, 2005), 97–98.

45. Richard F. Burton, *Zanzibar; City, Island, and Coast* (London: Tinsley Brothers, 1872), 2:223.

AFTERWORD

JOHN A. AGNEW

Science and its social geographies owe much to conversation and the connections that follow from it. Some years ago, David Livingstone and I fell into conversation about William Robertson Smith, a one-time Free Church of Scotland minister and theologian who, following a heresy trial, had been dismissed from his position as lecturer at Free Church College in Aberdeen but then had a second life at Cambridge University as Britain's most famous Orientalist. I knew little of this. I did know that Smith had written an article for the *Encyclopedia Britannica* on "Angels" that had been controversial at the time and subsequently among the largely rural congregations of the Free Church and, later, the breakaway Free Presbyterian Church of Scotland, with which my mother's family were affiliated. From their viewpoint, Smith had abandoned the Calvinist Westminster Confession of Faith for German-style biblical casuistry and the situating of religious belief in social life rather than simply in itself.

I have since discovered that Smith was precisely the type of "religious" intellectual whose interest in systematic textual analysis and precise investigation of sources has inspired David Livingstone's research into the boundaries between science and religion as a vantage point for understanding the intellectual history of the frequently mischaracterized, entirely "secularized" science of post-Enlightenment Europe and America.[1] Smith has, in fact, been the subject of two of Livingstone's essays and the topic of serious speculation as to his impact not just on biblical scholarship but also in influencing such diverse thinkers as Émile Durkheim, Marcel Mauss, Edward Evans-Pritchard, and Sigmund Freud. Livingstone has shown how Smith's proposals about the sociological basis to religion were much more threatening to the "Scottish Calvinist mind-set" than was evolutionary theory, even as

Smith himself sponsored a version of evolutionism that incorporated the very "textual and doctrinal history of Judaeo-Christianity."[2] Equally importantly, Smith, as with many other nineteenth-century figures of equivalent intellectual heft, did not abandon religious belief but saw his claims as "vindicating divine revelation."[3]

This brief diversion on the Robertson Smith story illustrates well the subtle and careful way in which David Livingstone contextualizes his subjects and their sources. It speaks, too, to the connections among science, religion, text, speech, and place that have characterized the geographical study of science and scientific culture—in which field Livingstone has left a distinct imprint. Smith was the social product of a range of venues, notably of German universities, outside the experience of those in Scotland who later criticized him. The move to Cambridge after his heresy trial liberated him from the intellectual defensiveness that necessarily afflicted a lecturer to ministers in training. Smith also made heroic attempts to put himself into the ancient settings of those biblical texts and other documents that were the subject of his investigations into "the religion of the Semites." Smith's intellectual pathway, briefly mapped here, is something of an archetype for the combined historicist and geographical rendering of science, religion, and intellectual history for which David Livingstone is discussed and celebrated in this book.

As the chapters in this volume show, David Livingstone's work has been at the forefront of what has been termed that "spatial" or "geographical turn" in science studies in general and historical studies of science more particularly.[4] This can, in one sense, be characterized as simply place- or venue-oriented, in claiming significance for distinctive geographical milieus such as towns, regions, and political territories in originating and offering distinctive readings of innovative scientific ideas and methods.[5] In another sense, it extends also to the movement and circulation of ideas over distance—science's ideas made mobile across space—and into new settings or venues where they emerge somewhere on a continuum between ready assimilation and reinvention, even reinterpretation. This emphasis is clear in Livingstone's work of 2014, for example, on the readings of Darwinism across the transatlantic world of the nineteenth century, but it is also rendered in more abstract terms in his seminal *Putting Science in Its Place* of 2003.[6]

As the editors make clear in their introduction, there has been considerable engagement in the history of science in recent decades over the relative merits of different geographical framings.[7] Manifestly, some fit better for some purposes than do others. Livingstone and I have elsewhere suggested that in addressing the problematic of "geographical knowledge" tout court, scholars should distinguish the role of particular sites of knowledge creation from the more general effects of geopolitical hierarchies on knowledge creation and dissemination, the specific impacts of place on perception and the possibility of new thinking, the geographical diffusion of hegemonic knowledge, and the circulation of ideas that makes the geography of reading knowledge different in different places.[8] In such a range of possibilities, it is clear that there is no single "spatial turn" but, rather, a number that can be matched to different analytic purposes. No single study could expect to invest in all of them. Equally, no single one of these spatialities supersedes or is of greater importance than the others. Place, space, mobility, and scale are mutually constitutive in this respect, relational not ranked. This is, I would observe, not always apparent when, for example, urban networks or territorial polities such as states or empires are accorded pride of place: this is scale in a strictly hierarchical sense and geography determined by political borders. Sadly, some versions of the spatial turn in contemporary intellectual history seem oblivious to the plurality of places, scales, and places, and to their variant uses in that intellectual history and in the geography of science.[9]

The chapters in this volume are each alert to the range of possibilities associated with the phrase "geographical" or "spatial turn." The book is organized around the concept of geographical scale—and a critique of that and other terms employed in the geographies of science—in order to make this approach apparent. Two of the chapters focus on the role of locales in the creation and reception of scientific and geographical knowledge. Robert J. Mayhew and Yvonne Sherratt's particular focus in their chapter is with "spatial hermeneutics"—namely, how generalizations made in specific texts, such as those of Thomas Malthus, draw from the particularities of everyday experience in specific locales, albeit with broader imperial connections. Theirs is a study of words and experiences shaped by place. In similar vein, Diarmid A. Finnegan uses the three visits to Belfast of Irish-born physicist John Tyndall, in 1852, 1874, and 1890, as his reputation swelled, to show how as the city changed in its sociopolitical complexion, the reception given to Tyndall also changed.

Three chapters privilege the national contexts in which scientific knowledge originated and circulated in the nineteenth century. Mark Noll uses the career of Henry Hotze in the southern United States to show how much his defense of slavery as an institution affected his scientific position on topics such as polygenesis, preadamitism, and evolution. Ronald L. Numbers provides a more general overview of American polygenism, tracing "evolutionary" lines of thinking through to the Scopes Trial in the 1920s and the spread of so-called creation science in the 1980s. He demonstrates how much the pre-Darwinian commitment to separate races has continued to inflect American discussions of evolutionism to the present day and has spread over space rather than remained simply associated with those places where it was long rooted. Nicolaas Rupke identifies and differentiates various "national schools" of evolutionary theory across the late nineteenth and early twentieth centuries, with a particular eye on Germany and beliefs there in structural evolutionary thinking. For him, the relative emphasis on functional response versus organic form in the British and the German cases, respectively, determined the reception to natural selection versus structural form as causal drivers of evolution.

The last section contains four chapters that focus on the global connections and spaces that were vital to the course of metrology and scientific exploration in the nineteenth century. Charles W. J. Withers traces the history of those international geographical meetings in the last third of the nineteenth century whose purpose, in part, was to decide upon the location of the prime meridian and lay out standardized time zones and an agreed metrological framework for the earth as a whole. The science in this case—a form of global physics, certainly of measurement proposed to a shared international standard—was a combination of national perspectives and the venue-specific influences operating at the meetings out of which a final and globally binding framework would emerge. Nuala C. Johnson uses the case of an Anglo-Irish woman, Charlotte Wheeler-Cuffe, to examine fieldwork in Burma as a heroic practice and a venue for the making of botany. Johnson brings into focus several themes central to nineteenth-century science generally: field experience and scientific reputation, the role of empire in providing opportunity (here for a married woman able to accompany her engineer husband), botany as a "polite" field of study in which the boundary between amateurs and professionals was always fuzzy and more open to women than some other sciences.[10] Vinita Damodaran addresses how much what later became the field of climatology and the burgeoning environmental determinism of the late nineteenth century had colonial roots. Contemporary debates about global warming, desertification, and deforestation are thus traceable in part to the ways in which colonial encounters and subsequent narratives have conditioned understandings of how the global climate system is based in local and regional processes rather than simply global in itself. Place matters materially as well as representationally. Finally, Dane Kennedy interrogates two early nineteenth-century official British expeditions to Africa, one to trace the Niger River, the other the Congo River. Both expeditions were failures on their own terms. But they illustrate some features of scientific exploration in Africa in general. One was the difficulty of topography, of overcoming geographical space, in accessing Africa's continental interior. Another was of adapting to local conditions. Heroic stories and narratives of science and the geographies of exploration as stories of unremitting success and conquest miss such complexities, sometimes deliberately.[11] They also miss the degree to which the "foreign" and commonly transient explorer relied on African assistance and assistants in facilitating access to the information that expeditions sought to collect. Scientific knowledge was wrested with difficulty and with a heavy dependence on indigenous sources and informants more than on the absorptive minds and not-always-resilient bodies of brilliant explorers.[12]

What generalizations may we draw from this collection? Noll's essay speaks to the complexities of the intersection between religion and evolutionary thinking that a focus on historical geography can highlight as well as to the intersections between locale and nation. Local and national politics frequently mediated between religious belief, on the one hand, and openness to scientific logic, on the other. This has been, of course, an important component of studies of science and religion in historical context. As I read this chapter, I thought of the danger, entirely avoided here as it is in David Livingstone's writing, of imposing contemporary notions of Protestant evangelical religiosity on interpretations of nineteenth-century Calvinism's reception of evolutionary thinking. Here, too, a moment of further biographical reflection is apposite. I finally appreciated why, many years ago, I objected to a point in a review by historian of science James Moore when he interpreted a 1987 work of David Livingstone's as problematic in terms of contemporary American Protestant Christianity, rather than situating it historically within a nineteenth-century Calvinism that, elsewhere, he had admitted was not narrowly committed to biblical inerrancy.[13] Moore may also have exaggerated the extent to which contemporary American evangelicalism tout court is devoted to biblical inerrancy and therefore somehow closed off from anything other than creationism. Numbers' chapter here discloses the complex origins and contemporary geographies of creationism in relation to nineteenth-century scientific and religious beliefs.

These observations, and the chapters by Noll and Numbers in particular, raise the broader question of the temporal-spatialities under which intellectual history and scientific knowledge operate. Even as ideas of the modern as cumulatively different from the past became increasingly popular by the late nineteenth century, symbolized as Withers shows by concerns over a universal time and a global metrology, a countermovement involved an emphasis on spatial-cultural diversity in which, quoting Claude Lévi-Strauss, "forms of civilization which we tended to imagine as 'succeeding one another in time' should rather be seen as 'spread out in space.'"[14] From this viewpoint, illustrated most clearly in the chapters of Finnegan, Rupke, and Numbers, persistent political and cultural differences between polities and localities condition scientific knowledge production and reception. Yet, as these authors also show, the contexts at issue do not remain fixed but are dynamic such that, for example, the Belfast Tyndall encountered in 1852 was fundamentally different culturally by 1890, not least because of the joint impact of the Great Irish Famine and its long-term effect on Catholic revival and the consequent politicization of non-Catholic denominations into an increasingly potent category of "Protestant."

Several of the chapters explore aspects hitherto relatively neglected with respect to intellectual history and the history of science. One is the imperial connections central to the chapters of Johnson and Kennedy. The former emphasizes the relative freedom for a Western woman to engage in scientific work in a colonial domain. The latter makes a strong case for the mediation of indigenous "assistants" in the entire corpus of colonial scientific expeditionary culture (with the exception of Antarctic science) in which heroic white men have long wrested knowledge from a recalcitrant wilderness and adversarial environments.[15] In both chapters, the role of mediators, go-betweens, and translators as well as institutions such as botanical gardens receive their due. This has become an important part of the rewriting of the history of science associated with the geographical turn.[16]

A second theme in terms of the geographical turn, per se, is that of geopolitical hierarchy. As Mayhew and Sherratt and Rupke, respectively, point out, albeit in different ways, ideas do not "win out" simply because of their veracity but do so because of the sociopolitical positioning of their proponents. Malthus's "universalizing ambition" was undoubtedly aided by his social position in the Britain of his day, however provincial his everyday existence might have been. It is also true that his writing was conditioned by the locale of his travels and his day-to-day lived experiences. Conversely, in the case of German structural evolutionism, this theory was always going to be at a relative intellectual and political disadvantage to Darwin's idea of natural selection, not simply on intellectual-empirical grounds but because Darwin had Britain's global reach on his side. Yet in recent years, the emphasis on the constraining and directing impact of inherited physical forms has undergone a revival in biology even as natural selection sensu stricto has been subject to requestioning on decidedly noncreationist grounds.[17] Geopolitical hierarchies, we should not forget, are undermined and change over time.

The third geographical theme is that of mobilities of knowledge. Circulation, as Livingstone has dubbed the ways ideas travel from place to place, was absolutely vital to the engagements with scientific knowledge that he has traced across his many publications on Darwinism in particular and science more generally. Recently, he has shown how much ideas of environmental determinism resurrected from the early twentieth century have been recycled and recirculated to justify, for example, apocalyptic claims about the links between climate change and warfare without any sort of human-institutional mediation.[18] Several of the chapters in this collection also point out that the geography of knowledge is not just about where something is written or who writes it but also how what is written is read and received elsewhere. This is arguably one of the most important contributions to have arisen from study of the history of science read geographically. In his chapter, Numbers lays stress on how much old ideas do not die but go to earth, so to speak, and then come back to life when they travel elsewhere. Rupke mentions how Darwin traveled textually, so to speak, to the former Soviet Union but was institutionalized there at least for a while without that influence of Malthusian political economy that came to predominate in the Western "Darwinian synthesis" of the mid-twentieth century. Ideas are adapted as they travel. Mobility over space can make a difference to what an idea is, or is thought to be. What Wendy Griswold has called "the reading classes" thus take on a role in the history of science from which they were hitherto excluded by the almost exclusive emphasis on the production of knowledge.[19] Arguably, therefore, rather than just the sense of location or place as the predominant sense of geography informing the geographical turn, it is the role of knowledge mobility in dialectical dance with the specificities of place that represents the main contribution of the overall approach.[20]

This collection does a very fine job of extending in varied ways many of David Livingstone's contributions to the historical geography of science. Taken together, the chapters capture much of the same attention to the contingent intersection of sites, venues, practices, performance, and audiences that Livingstone and Withers put together in 2011.[21] These essays also build upon that work. Without being prescriptive as to focus or method, they point to the possibilities and rewards of future studies in the geographies of science, and not just for the nineteenth century. They also much more explicitly address the various types of spatiality involved in the production and reception of knowledge, particularly in the making of science. If, as noted, science and its geographies are strongly biographical, it is worth considering that the approaches adopted by these chapters and in the geographies of science as a whole can be turned on Livingstone and his work. When you hail from a periphery that those in the metropolis scarcely recognize let alone understand, there is a compunction to put them straight. This is partly about gaining recognition for your place and its people. It can also lend itself, however, to a broader sense that knowledge production and circulation are not naturally self-organizing and distributing or reducible to presumptions about immediate availability and common readings. They are, rather, the outcome of systematic interests that reflect geographical structures such as core-periphery, urban hierarchies, histories of colonialism and settlement, the historic sedimentation of political-economic and religious distinctions, and the active marginalization of those with distinctive and different religious and social identities. That David Livingstone has lived most of his life in Belfast and has had a long affiliation with Queen's University Belfast seems not simply incidental to the geography of intellectual history that he has crafted over the past forty or so years.

We may then look to matters of locale and to the national to help explain his global reputation. I would single out three influences that have probably played some part in affecting the particular path to intellectual agency that David Livingstone has followed. The first is his own background as an Irish Presbyterian, an affiliation he would be quick to identify as needing equal emphasis on both words. This has been one of the inspirations for his intensive and extensive explorations of the intersections between Presbyterianism in its various nineteenth-century manifestations across the transatlantic world and beyond, on the one hand, and the reception and interpretations of evolutionary biology, on the other. But its influence extends also into his broader perspective on knowledge creation and circulation. The self-image of Presbyterianism has been of religious belief that is not simply enthusiastic in the sense, say, of Pentecostalism or ritualistic in the sense, say, of Catholicism and Anglicanism but rigorously grounded in an intellectual process of self-discovery and correction.

The second is the fact of living in Ireland and being subject to English stereotypes of a peculiarly environmentally determinist character. One need not endorse wholly Estyn Evans's *The Personality of Ireland* to acknowledge his point when, speculating on the fate of Ireland as reflecting in English eyes, to quote one authority, how "the people of England, subject to the climate that now afflicts Ireland, would immediately begin to develop all those shortcomings which they are now so ready to find in the Irish."[22] Livingstone's well-argued hostility to environmental determinism in its various guises is not reducible to his place of residence, but it certainly could be conditioned by it.

Lastly, David has lived his life in and around a city where the population literally wears its identities on its sleeves. This is not just a question of the religious-nationalist sectarianism and political violence that Belfast has come to symptomatize in certain quarters. It is more a question of how living there sensitizes you to the political and cultural mediations that affect the practices and performances that constitute social life—and scientific life—and the extent to which these are grounded in definite locales and venues. What is evident there is more hidden away in places with less fervent identity struggles. Yet at the same time, people from Belfast are intimately connected to family and former neighbors in a scattered global diaspora. There is an immediacy to the transnational and to the global that is missing, perhaps, in more self-contained and prosperous communities. Anthropologists of Ireland Thomas Wilson and Hastings Donnan quote geographer Jean Gottmann as capturing the politics of emplacement at work: "The main partitions observed . . . [by] . . . men are not those in the topography or in the vegetation, but those that are in the minds of men."[23]

I recall David Livingstone talking fondly of three Irish writers and intellectuals whose differences and overlaps say something about how his own worldview formed and has influenced what he does. One was the great late poet and Queen's graduate Seamus Heaney. The second was the retired-to-Donegal philosopher of science from the University of Notre Dame in the United States, Ernan McMullin. The third, Oxford don and writer C. S. Lewis, grew up down the street from where David Livingstone lives. These three men do not fit neatly into a single box. That speaks for David Livingstone, too. The same positive vibe of Belfast singer Van Morrison also inflects Livingstone's outlook. The joy, soul, and humor of Belfast inform the man. At the heart of Livingstone's work is what might appear as a paradox to the linear-minded: that religion and science are not simple opposites and can occupy common ground.[24] Umberto Eco made this point when writing of Andrew Dickson White's two-volume treatise *History of the Warfare of Science with Theology in Christianity* that "his aim is to list every instance in which religious thought impeded the advancement of science, but as he is an informed and honest man he cannot conceal the fact Augustine, Albertus Magnus, and Aquinas knew very well that the earth was round. He adds, however, that to sustain this idea they had to combat dominant theological thought. But dominant theological thought was represented, in fact, by Augustine, Albertus, and Aquinas, who thus had to combat no one."[25] Oppositions only go so far, as anyone in Belfast can tell you.

This volume illustrates the vitality intrinsic to the geographical study of science. The chapters demonstrate the importance of thinking spatially, whether about science, popular scientific culture, intellectual history, religious adherence, or racism. We can hope, just as it builds upon and extends from earlier work, that this volume will, in turn, prompt further research. The terminology and the methods of analysis employed in the spatial study of science and of intellectual history have now a consistency and scholarly presence that, years ago, they did not enjoy. That this is so is not to determine in advance the subjects, the places, or the means by which science in the past was read, or spoken, or listened to or in other ways performed and so should be interrogated by modern researchers. It is to invite yet further scrutiny of science's political and cultural dimensions and of its cognitive content and, perhaps, as do these chapters, to acknowledge the role of David Livingstone's scholarship in stimulating such work.

Notes

1. One of David's earliest forays in this direction is the landmark essay "Science and Religion: Foreword to the Historical Geography of an Encounter," *Journal of Historical Geography* 20 (1994): 367–83.

2. David N. Livingstone, "Public Spectacle and Scientific Theory: William Robertson Smith and the Reading of Evolution in Victorian Scotland," *Studies in the History and Philosophy of the Biological and Biomedical Sciences* 35 (2004): 1–29. On Robertson Smith and his influence on the budding social sciences, see, among others, T. O. Beidelman, *W. Robertson Smith and the Sociological Study of Religion* (Chicago: University of Chicago Press, 1974), and Alexandra Maryanski, "The Birth of the Gods: Robertson Smith and Durkheim's Turn to Religion as the Basis to Social Integration," *Sociological Theory* 32 (2014): 352–76.

3. David N. Livingstone, "Finding Revelation in Anthropology: Alexander Winchell, William Robertson Smith, and the Heretical Imperative," *British Journal for the History of Science* 48 (2015): 435–54.

4. For an excellent overview, see, for example, David N. Livingstone, "Landscapes of

Knowledge," in *Geographies of Science*, ed. Peter Meusburger, David N. Livingstone, and Heike Jöns (Dordrecht: Springer, 2010), 3–22.

5. See, for example, Simon Naylor, "Introduction: Historical Geographies of Science—Places, Contexts, Cartographies," *British Journal for the History of Science* 38 (2005): 1–12; Diarmid Finnegan, "The Spatial Turn: Geographical Approaches in the History of Science," *Journal of the History of Biology* 41 (2008): 369–88; Richard Powell, "Geographies of Science: Histories, Localities, Practices, Futures," *Progress in Human Geography* 31 (2007): 309–30.

6. David N. Livingstone, *Dealing with Darwin: Place, Politics, and Rhetoric in Religious Engagements with Evolution* (Baltimore: Johns Hopkins University Press, 2014); David N. Livingstone, "Debating Darwin at the Cape," *Journal of Historical Geography* 52 (2016): 1–15; David N. Livingstone, *Putting Science in Its Place: Geographies of Scientific Knowledge* (Chicago: University of Chicago Press, 2003).

7. See, among others, James Secord, "Knowledge in Transit," *Isis* 95 (2004): 654–72; Charles W. J. Withers, "Place and the 'Spatial Turn' in Geography and History," *Journal of the History of Ideas* 70 (2009): 637–58; and Jouni-Matti Kuukkanen, "Senses of Localism," *History of Science* 50 (2012): 477–500.

8. John A. Agnew and David N. Livingstone, "Introduction," in *Sage Handbook of Geographical Knowledge*, ed. John A. Agnew and David N. Livingstone (London: Sage, 2011), 1–17. For an innovative effort at combining such relational and territorial geographies of science, see, for example, Richard McMahon, "Networks, Narratives and Territory in Anthropological Race Classification: Towards a More Comprehensive Historical Geography of Europe's Culture," *History of the Human Sciences* 24 (2011): 70–94.

9. See, for example, Peter J. Taylor, Michael Hoyler, and David M. Evans, "A Geohistorical Study of 'the Rise of Modern Science': Mapping Scientific Practice through Urban Networks, 1500–1900," *Minerva* 46 (2008): 391–410, and Lauren Benton, "Afterward: The Space of Political Community and the Space of Authority," *Global Intellectual History* 3 (2018): 254–65.

10. On the connections between botany and gender, and botany as a "safe" science for women at least in England and for particular status groups, see Ann B. Shteir, *Cultivating Women, Cultivating Science: Flora's Daughters and Botany in England, 1760–1860* (Baltimore: Johns Hopkins University Press, 1996).

11. This point is examined, for many explorer-authors in the first half of the nineteenth century, in Innes M. Keighren, Charles W. J. Withers, and Bill Bell, *Travels into Print: Exploration, Writing, and Publishing with John Murray, 1773–1859* (Chicago: University of Chicago Press, 2015).

12. On this point and with reference to East African exploration in the mid-nineteenth century, see Stephen J. Rockel, "Decentering Exploration in East Africa," in *Reinterpreting Exploration: The West in the World*, ed. Dane Kennedy (Oxford: Oxford University Press, 2013), 172–94. On explorers' physical and mental frailty in African exploration, see Johannes Fabian, *Out of Our Minds: Reason and Madness in the Exploration of Central Africa* (Berkeley: University of California Press, 2000).

13. The work in question is David N. Livingstone, *Darwin's Forgotten Defenders: The Encounter between Evangelical Theology and Evolutionary Thought* (Edinburgh: Scottish Academic Press, 1987). On the review, see James R. Moore, "An Equivocal

Heritage," *Science* 40 (1988): 1049–50. On the admission of Calvinism's openness to Darwinism, see James R. Moore, *The Post-Darwinian Controversies: A Study of the Protestant Struggle to Come to Terms with Darwin in Great Britain and America, 1870–1900* (Cambridge: Cambridge University Press, 1979).

14. Quoted in François Hartog, *Regimes of Historicity: Presentism and Experiences of Time* (New York: Columbia University Press, 2015), 14.

15. See, for example, Richard C. Powell, *Studying Arctic Fields: Cultures, Practices, and Environmental Sciences* (Montreal: McGill-Queens University Press, 2017), for a study of geographical and ethnographic science in high latitudes. For a study of glaciological science at high altitude, see Philip W. Clements, *Science in an Extreme Environment: The 1963 American Mount Everest Expedition* (Pittsburgh, PA: University of Pittsburgh Press, 2018), and Danielle K. Inkpen, "The Scientific Life in the Alpine: Recreation and Moral Life in the Field," *Isis* 109 (2018): 515–37.

16. See, for example, Miles Ogborn, " 'It's Not What You Know. . .': Encounters, Go-betweens and the Geography of Knowledge," *Modern Intellectual History* 10 (2013): 163–75, and Kapil Raj, "Go-betweens, Travelers, and Cultural Translators," in *A Companion to the History of Science*, ed. Bernard Lightman (Oxford: Wiley-Blackwell, 2016), 39–57.

17. On the former, see Stephen J. Gould and Richard Lewontin, "The Spandrels of San Marco and the Panglossian Paradigm: A Critique of the Adaptationist Programme," *Proceedings of the Royal Society of London*, B, 2005 (1979): 581–98. On the latter, see Keith D. Bennett, "The Chaos Theory of Evolution," *New Scientist* 208 (2010): 28–31.

18. David N. Livingstone, "The Climate of War: Violence, Warfare, and Climatic Reductionism," *WIREs Climate Change* 6 (2015): 1–8.

19. Wendy Griswold, *Regionalism and the Reading Class* (Chicago: University of Chicago Press, 2008). For one discussion on the geography of scientific reading in the nineteenth century, see James A. Secord, *Victorian Sensation: The Extraordinary Publication, Reception, and Secret Authorship of* Vestiges of the Natural History of Creation (Chicago: University of Chicago Press, 2000).

20. On this, see, for example, Heike Jöns, Peter Meusburger, and Michael Heffernan, eds., *Mobilities of Knowledge* (Dordrecht: Springer, 2017).

21. David N. Livingstone and Charles W. J. Withers, eds., *Geographies of Nineteenth-Century Science* (Chicago: University of Chicago Press, 2011).

22. E. Estyn Evans, *The Personality of Ireland: Habitat, Heritage and History* (Dublin: Lilliput Press, 1992), 20.

23. Jean Gottmann, "Geography and International Relations," *World Politics* 3 (1951): 164, quoted in Thomas M. Wilson and Hastings Donnan, *The Anthropology of Ireland* (Oxford: Berg, 2006), 115.

24. David Livingstone and John Hedley Brooke discussed this topic publicly at Queen's University Belfast on November 19, 2018: https://theconversation.com/war-between-science-and-religion-is-far-from-inevitable-106477.

25. Umberto Eco, *Serendipities: Language and Lunacy* (New York: Columbia University Press, 1999), 9.

INDEX